New Advances in Semiconductors

Edited by Alberto Adriano Cavalheiro

Published in London, United Kingdom

New Advances in Semiconductors
http://dx.doi.org/10.5772/intechopen.97948
Edited by Alberto Adriano Cavalheiro

Contributors
Peter Cendula, Maan Hameed, Adil-Gerai Kussow, Edosa Tasisa Jira, Daniela Munteanu, Jean-Luc Autran, Pantea Nadimi Goki, Luca Poti, Antonio Tufano, Fabio Cavaliere

First published in London, United Kingdom, 2022 by IntechOpen
IntechOpen is the global imprint of INTECHOPEN LIMITED, registered in England and Wales, registration number: 11086078, 5 Princes Gate Court, London, SW7 2QJ, United Kingdom

British Library Cataloguing-in-Publication Data
A catalogue record for this book is available from the British Library

Additional hard and PDF copies can be obtained from orders@intechopen.com

New Advances in Semiconductors
Edited by Alberto Adriano Cavalheiro
p. cm.
Print ISBN 978-1-80355-681-9
Online ISBN 978-1-80355-682-6
eBook (PDF) ISBN 978-1-80355-683-3

For EU product safety concerns:
IN TECH d.o.o., Prolaz Marije Krucifikse Kozulić 3, 51000 Rijeka, Croatia,
info@intechopen.com or visit our website at intechopen.com.

Meet the editor

Alberto Adriano Cavalheiro is an associate professor at the State University of Mato Grosso do Sul (UEMS), Brazil, where he works as a permanent lecturer in the Graduate Program in Natural Resources and coordinates the LIMAN materials laboratory. He holds a bachelor's, master's, and a doctorate in Chemistry and a Licentiate of Science. He completed two postdocs with research in semiconductors and catalysts. He works in teaching, research, and extension, with a focus on chemistry and material and environmental sciences. He also coordinates the chemistry area of the Teaching Initiation Program at UEMS. He has numerous funded projects, articles in peer-reviewed journals, and book chapters to his credit. Dr. Cavalheiro has mentored several undergraduate and graduate students.

Contents

Preface

Semiconductor materials have a long design and manufacturing chain before their use in all types of electronic products, which also need to incorporate various aspects of circuit logic to adapt to current and future needs. Revolutions in the automotive, media and Internet markets depend on this technological synergy, as do other sectors such as banking and financial markets, telemedicine, e-commerce, industrial automation, smart homes, and agriculture, all of which are expected to benefit from these advances.

New power-generation processes and the development of more energy-efficient devices and equipment directly or indirectly involve new advances in semiconductors. Quality of life has greatly improved with the mass production of more affordable devices for individual communication, which directly depends on the reduction of semiconductor costs. Therefore, more efficient and cheaper semiconductor materials can partially solve the problem of the imbalance between energy generation and consumption by individuals and communities.

Semiconductor materials are the basis of photovoltaic cells for on-site electricity generation. They are also used in water purification processes and hydrogen fuel generation. Thus, the field of semiconductor research is interdisciplinary, including chemistry, physics, engineering, and medicine, offering technological solutions in various fields.

New Advances in Semiconductors offers an updated review of semiconductors, including strategies in the field of circuit logic and experiments on some semiconductor materials. It discusses important aspects of the performance of semiconductors, such as environmental parameters of operation optimization, namely, the influence of operating temperature, light incidence, electric current, magnetic field, and even radioactivity. Many devices are influenced by variations of these experimental parameters, from simple diodes and transistors to complex integrated circuits and solar cells.

This book contains six chapters, divided into two sections. Section 1, "Calculations and Simulations in Semiconductors," includes four chapters that discuss several aspects of the mechanisms of action in systems and devices based on semiconductors. They present proposals to obtain greater energy savings in an operational environment and examine how the interfacial mechanisms between semiconductors and electrolytes can be better understood for the continuous improvement of devices in future projects.

Chapter 1 shows how considerations about individual particles introduce additional difficulties in quasi-macroscopic dielectric function calculations and proposes the exclusion of the single-particle effective field model in this type of calculation based on dielectric relaxation matrices. The author proposes that the use of a polyelectronic wave function, as a sufficient condition for the calculations, has the advantage of allowing considerations about the interactions of the electron-hole pair to be disregarded, greatly facilitating the calculations.

Chapter 2 presents a new analytical model for calculating performance in amplitude-modulated systems, which can be very useful for semiconductor optical amplifiers, considering the lower and upper limits of performance. These calculations can be useful for the development of integrated photonic circuit designs that use amplitude-modulated systems, such as optical amplifiers and interconnects. The model is validated through numerical and experimental simulations, including examples of photonic integrated circuit sizing, and shows how the parameters associated with amplitude modulation can be adjusted for application in semiconductor optical amplifiers.

Chapter 3 discusses how the use of a new circuit design based on Tri-State Buffer connection technology can lead to greater efficiency in electronic circuits. The project shows that there are energy savings if a submodule with two logic routines based on ASIC design and Verilog HDL language is inserted into the logic circuit diagram. Thus, the circuit presents greater energy savings due to the reduction of dynamic and total power. These are relevant mathematical approaches, as they demonstrate how energy savings are verified as a function of dissipation reduction. A performance evaluation is also provided for validation in terms of energy dissipated in connections with Tri-State Buffer technology.

Chapter 4 provides an introduction to charge transport at the interphase junction between solid semiconductors and liquid electrolytes. It demonstrates how the changes occur whether the system is illuminated or not, with basic considerations on the use of semiclassical charge transport equations in the semiconductor used as an electrode. The chapter also presents the behavior of the ideal interface between the semiconductor and the electrolyte at a broad level, describing the interactions as a function of the redox reactions in the electrolyte and the energy relations of charge transfer. The proposed system is also interpreted from the point of view of the kinetics of the electron-hole pair recombination during irradiation. Propositions are also made on how the semiconductor surface can be occupied to explain the behavior of the photocurrent-voltage curves.

Section 2, "Semiconductor Materials," contains two chapters. These chapters discuss two important environmental parameters in semiconductor operation. Chapter 5 describes the dependence of electrical resistivity as a function of temperature for dilute magnetic semiconductors. It also presents a specific case of electrical resistivity reduction for a manganese-doped gallium arsenide p-type semiconductor and examines how the behavior of the curves changes as a function of temperature from 0K to 300K. It is shown that, in addition to the temperature dependence, the electrical resistivity of this type of magnetic material also strongly depends on the concentration of dopants in the structure. Considerations on mechanisms of how electrical resistivity considerably affects state density and energy gap are also made.

Chapter 6 discusses the properties and consequences of the response in electronic components and circuits based on semiconductor materials exposed to neutron radiation. It presents the behavior of several semiconductors based on chemical elements of Group IV and Groups III and V of the periodic table regarding two types of neutron radiation produced by reactions involving deuterium as well as deuterium and tritium. Reaction rates and classification of neutron-induced byproducts are presented, according to their nature and energy.

Please note that this book can be read in or out of sequential order, as there is no progression of basic knowledge level between chapters or sections. We thank each of the chapter authors for bringing us the most current issues in this field of knowledge. This book is a useful resource for students, researchers, and other interested readers.

Alberto Adriano Cavalheiro
Interdisciplinary Laboratory of Advanced Materials of Navirai (LIMAN),
State University of Mato Grosso do Sul (UEMS),
Navirai-MS, Brazil

Section 1

Calculations and Simulations in Semiconductors

Chapter 1

Many-Electron Problem in an Atomic Lattice Reduced Exactly to Two-Particle *Pseudo-Electron* Excitations: Key to Alternative First-Principles Methods

Adil-Gerai Kussow

Abstract

Prediction of properties of solids (semiconductors) is based almost entirely on the first-principles methods. The first principles theories are far from being perfect and new schemes are developing. In this study, we do not follow the traditional one-particle-in-effective-field concept. Instead, all Coulomb interactions between particles are treated in their original form, i.e., particle-particle discrete interactions. Two-particles Coulomb excitations theory in a crystal lattice is proposed, along with a method for calculations of physical measurables. Most important, the relevant particles are *not* electrons but *pseudo-electrons* with both the Coulomb interaction mode and the effective mass different from those of electrons. The unitary transformation represents the many-body system as an ensemble of two-pseudo-electron excitations *without neglection* of the terms in a Hamiltonian. The many-particle wave function, being derived in a non-trivial two-particle form, ensures a full description of exchange-correlation and screening effects, for *both ground and excited states*. As an example, the energy of a many-electron system and the quasiparticle energies are expressed in an elegant integral closed-form and compared with the Density Functional Theory. The proposed scheme possibly opens a new route toward the numerical evaluation of properties of many-particle systems.

Keywords: solid state, quantum mechanics, many-electron problem, first principles, properties of semiconductors

1. Introduction

Since the creation of Quantum Mechanics (QM), the theories for the description of many-electron systems, with *N* electrons, have been developed for decades [1, 2]. The most elaborated applications of many-body theories are the first-principles calculations which are used virtually in all fields of modern physics [3]. These calculations became a tool that provides priceless information to describe both the undisturbed atomic configuration and the response of the system to an external perturbation. In terms of formalisms, the many-electron theories belong to several groups: 1. Perturbation-based theories [4]. These methods utilize a small α parameter expansion, which, often, is the ratio of the electron–electron Coulomb interaction energy to the total energy, $\alpha = \|H_{\text{int}}/H_0\|$. 2. Green function-oriented theories are generally more powerful than perturbation schemes, since they are not necessarily based on a small parameter. 3. Density functional-based schemes (DFT) [5–8] explore a semiclassical concept of an electron gas [9]. 4. The coupled cluster theories [10] find solutions of the Schrodinger equation for many-electron problem without assumptions of DFT theory (e.c. electrons gas approximation). The goal of the cluster theories is the calculation of the wave function for the many-body system, $\Psi\left(\vec{r}_1, \vec{r}_2, \ldots \vec{r}_N\right)$, which depend on N coordinates, $\vec{r}_i$, of electrons. To treat the exchange-correlation effects, the coupled cluster techniques utilize the antisymmetric Slater determinant, D. The elements of D are the one-electron wave functions, $\varphi_i\left(\vec{r}_i, t\right)$, derived within the effective-field approximation. The wave functions, $\varphi_i\left(\vec{r}_i, t\right)$, are calculated assuming two or three electron excitations (Coulomb interactions between electrons), and D provides the many-electron wave function, $\Psi\left(\vec{r}_1, \vec{r}_2, \ldots \vec{r}_N\right)$. Different methods [10–13], which include the parametrization and variational principles, allow to derive the ground and the excited states of the system. The coupled cluster approaches often demonstrate the superior accuracy of calculations if compare with DFT schemes, and these methods were successfully applied in different fields ranging from Chemistry to Nuclear Physics [10].

Despite differences, all first-principles theories have a common foundation—the concept of *one-particle in an effective-field* concept. This concept leads to the description of a many-particle system in terms of one-electron wave functions, $\varphi_i\left(\vec{r}_i, t\right)$, which are wave functions of free electrons, corrected due to the presence of other particles. The main difficulty here is that all N electrons interact with each other, with a large number of interactions $N(N-1)/2 >> 10$. Consequently, the first-principles schemes consider a restricted number of Coulomb interactions between the electrons only, and the rest of the interactions are omitted. The numeric schemes usually select combinations of 2–4 electrons, and the higher interaction orders are ignored. If Coulomb interactions are considered *exclusively* between couples of electrons (and the interactions of higher order are neglected), these schemes are called two-particle methods. Consequently, only the approximation solutions are possible and different sophisticated methods are utilized to increase the accuracy [3–13].

In this study, we look at the many-body problem from a very different angle. First, we try to find the answer to the following question: Is it possible to reduce the many-electron problem to the *exactly two-particle* scenario when each particle interacts by a Coulomb force solely with one another particle? The motto here is the main principles of Quantum Mechanics which always brings into focus the most elementary levels of any effect. Obviously, in terms of interactions between particles, the interaction between *two* particles is an elementary interaction, and all other interactions, are superpositions of two-particle interactions. Hence, one may ask if it is possible to canonically transform the many-body Hamiltonian, $\hat{H}$, to a form that involves *exclusively* two-particle interactions, and all other excitation orders are not presented?

If this question is positively answered, the relevant particles in canonically transformed Hamiltonian obviously *are not electrons* and should have different properties.

Here, we demonstrate that such canonical transformation, $\tilde{T}$, exists. Consequently, we introduce new particles or *pseudo-electrons* that adequately represent the canonically—transformed many-body system. The two-particle excitations describe the mutual scattering of coupled pseudo-particles in a periodic crystal field that obeys the Schrodinger equation. Note, that these excitations are different from both quasi-electrons and excitons which describe the interaction between an excited electron and a hole.

In other words, we show here that many-electron problems can be *exactly* expressed as an ensemble of two-particle excitations, without any truncation or omissions in Hamiltonian. Since the two-particle problem was extensively studied in many QM treatments and can be solved by different methods [14], our approach promises an elegant solution for the many-electron problem. Strangely, in a sea of literature on the many-body problem, we could not find analogous canonical transformation. All two-particle theories in the many-body problem, e.c. double cluster methods [10–13] or DFT [5–8], consider *electrons* in a truncated Hamiltonian with omitted high order electronic interactions. Another example is the *geminals* in Quantum Chemistry [15, 16] or two-particle coordinate (spin) wave functions which represent a generalization of one-electron orbitals accounting for intra-orbital correlation effects. Again, the geminal theories assume that the relevant particles are *electrons* and not pseudo-electrons. Note, that our pseudo-electrons are *very different* from electrons in terms of their properties, which are their effective mass and the Coulomb mode of interaction. Moreover, the total number of pseudo-electrons, $N(N-1)$, is different from the total number, N, of electrons.

Consequently, we believe that our approach is very different from existing many-body theories. The numerical justification of our scheme requires special extensive work and is surely out of a scope of a current study. Meantime, we hope to provide such numerical validations later. The theory utilizes the following methods: special unitary trans-formation of many-electron Hamiltonian, $\hat{H} \rightarrow \tilde{\hat{H}}$, two-particle Green function formalism for the calculation of the two-particle excitations, and combinatorics applied to Slater determinant for the derivation of a fermionic many-body wave function, $\Psi\left(\vec{r}_1, \vec{r}_2, \ldots .. \vec{r}_N\right)$. All parts of the approach are within traditional QM formalism [17, 18], and no other supporting approximations were utilized. We have found out, that our efforts are extremely rewarding in terms of results. They lead to general expression, in a closed form, for the main object of QM—many-electron fermionic wave function, $\Psi\left(\vec{r}_1, \vec{r}_2, \ldots .. \vec{r}_N\right)$. Just as an example, we derive from Ψ the electronic lattice energy functional which is the source for the calculation of quasi-particles (QP) energies, band structures, and optical spectra. These spectra are often obtained within Hedin's GW approach [19] based on one-body single-particle Green's function formalism. The QP spectra are required to obtain the optical spectra, which are often calculated from the Rohlfing-Louie electron–hole excitation model [20, 21], or the many-body perturbation Green function method. In a contrast, our approach allows us to calculate QP spectra, and hence, the optical spectra as well, based solely on the many-electron fermionic wave function, Ψ.

2. Unitary transformation of a Hamiltonian

We consider the periodic crystal lattice with cyclic boundary conditions, which has totally N_0 electrons per unit cell. We are interested in quantum states of $N < N_0$

electrons of a lattice, assuming, that the rest of $(N_0 - N)$ electrons, and core ions, can be replaced by an effective space-dependent pseudo-potential $V_{ext}\left(\vec{r}\right)$. Note, that the all-electron problem, $N = N_0$, when the pseudo-potential, $V_{ext} \equiv V_{ION}$, is solely due to atomic nuclei, is covered within the same theory. The pseudo-potential, $V_{ext}\left(\vec{r}\right)$, generates the set, $\{i\}$, of the single-electron states, with indexes, i and the wave vector, $\vec{k}_i$. Some of the states are occupied with N electrons of our interest, having coordinates, $\vec{r}_i$ and spin, σ_i. Each singe-electronic state can be occupied by two electrons, with opposite spins, by one electron, or is an empty state. The initial electronic state of a crystal is a set, $\{i\}$, of single-electronic states of our choice, with suppressed Coulomb interactions within the group of N electrons. The single-electronic states, with an index i and a wave vector $\vec{k}_i$, are described by one-electron ortho-normalized wave functions, $\varphi^0_{i\vec{k}_i}\left(\vec{r}_i\right)$, which are supposed to be known or previously calculated (e.c. these states can have the eigenfunctions for the pseudo-potential widely used in first-principles calculations). Note, that the initial electronic state of a many-electron system can be a ground state *or any excited* state of our interest. Each single-electronic state, with an index, s, if double occupied with electrons having opposite spins, splits into two degenerated electronic states. Next, based on the initial electronic state of a crystal, we turn on the Coulomb interactions between N electrons. Our goal is a calculation of the actual quantum states of the lattice, or the many-particle wave function of the system, Ψ, with all electron–electron Coulomb interactions included. This formulation of a many-body problem is a traditional one in first-principles calculation schemes and analytical many-body theories. The nonrelativistic Hamiltonian, $\hat{H}$, for the many-electron problem is given by:

$$\hat{H} = \sum_i \left(-\frac{\hbar^2}{2m} \nabla_i^2 + V_{ext}\left(\vec{r}_i\right)\right) + \sum_{i,l>i} \frac{e^2}{|\vec{r}_i - \vec{r}_l|} \tag{1}$$

where the last term describes the Coulomb interactions between electrons. For simplicity, we omit here the spin-orbital and other high-order interactions. To derive two-electron Coulomb excitations, one needs to apply to $\hat{H}$ a special unitary transformation. One can see that, in Hamiltonian (1), the sums are running over both one integer index (first and second terms) and two integer indexes (third term). This inconsistency or imparity in the number of indexes can be fixed with help of algebraic equality:

$$\sum_{i=1}^{N} a_i = \frac{1}{(N-1)} \sum_{s=1}^{N} \sum_{r>s} (a_s + a_r) \tag{2}$$

where a_i is a member of a set (for example, if $N = 3$, then Eq. (2) reads:

$$a_1 + a_2 + a_3 = (a_1 + a_2)/2 + (a_1 + a_3)/2 + (a_2 + a_3)/2 \tag{3}$$

The unitary transformation, $U^+ = U^{-1}$, of the Hamiltonian, $\hat{H}$, is equivalent to re-grouping terms of $\hat{H}$, based on equality (2). This transformation, being applied to one-indexed terms of $\hat{H}$, makes all terms of the Hamiltonian consistent in the number of indexes, which are now two. As a result, the transformed Hamiltonian $\tilde{\hat{H}} = U\hat{H}U^{-1}$

becomes the sum of two-particle Hamiltonians, $\tilde{\hat{H}}_{il}\left(\vec{r}_i, \vec{r}_l\right)$, which describe the two-particle excitations, with particle indexes, $\{i, l > i\}$:

$$\tilde{\hat{H}} = \sum_{i,l>i} \hat{H}_{il}\left(\vec{r}_i, \vec{r}_l\right) \tag{4}$$

$$\hat{H}_{il}\left(\vec{r}_i, \vec{r}_l\right) = \frac{1}{(\mathrm{N}-1)}\left(-\frac{\hbar^2}{2m}\nabla_i^2 - \frac{\hbar^2}{2m}\nabla_l^2 + V_{ext}\left(\vec{r}_i\right) + V_{ext}\left(\vec{r}_l\right)\right) + \frac{e^2}{|\vec{r}_i - \vec{r}_l|} \tag{5}$$

One can see, from Eq. (5), that the Hamiltonians $\hat{H}_{il}\left(\vec{r}_i, \vec{r}_l\right)$ can be written as:

$$\hat{H}_{il}\left(\vec{r}_i, \vec{r}_l\right) = -\frac{\hbar^2}{2\tilde{m}}\nabla_i^2 - \frac{\hbar^2}{2\tilde{m}}\nabla_l^2 + \tilde{V}_{ext}\left(\vec{r}_i\right) + \tilde{V}_{ext}\left(\vec{r}_l\right) + \frac{e^2}{|\vec{r}_i - \vec{r}_l|} \tag{6}$$

with both the renormalized mass of electrons, $\tilde{m} \to (N-1)m$, and the renormalized effective potential, $\tilde{V}_{ext}\left(\vec{r}\right) \to V_{ext}\left(\vec{r}\right)/(N-1)$. This renormalization, does not affect Coulomb interactions, $H_{\mathrm{int}}^{il} = e^2/|\vec{r}_i - \vec{r}_l|$, and $\tilde{e} = e$. In the following text we will call the electrons with renormalized mass as *pseudo-electrons*, and the renormalized pseudo-potential as an *effective pseudo-potential.* Each original electron, with index s, in an effective pseudo-potential, $\tilde{V}_{ext}\left(\vec{r}_s\right)$, on energy level, E_s, after the unitary transformation, splits into a group of $(N-1)$ pseudo-electrons, each having the same space-spin coordinates, $\left\{\vec{r}_s, \sigma_s\right\}$. The general coordinates of the pseudo-electrons, $\xi_S = \left\{\vec{r}_S,\ \sigma_S\right\}$, include both the space coordinates, $\vec{r}_S$, and the spin coordinates, σ_S. The total number of pseudo-electrons, $2C_N^2 = N^2 - N$, is larger than N, and each pseudo-electron interacts with solely one pseudo-electron from the other group. **Figure 1**, or an example of four electron system, illustrates these points more clearly. Note, that the pseudo-electrons, having the same index s, interact only with pseudo-electrons having the *different index*, s', with no interactions between pseudo-electrons within the same group. In a transformed lattice, the system of N electrons in an effective pseudo-potential, $V_{ext}\left(\vec{r}\right)/(\mathrm{N}-1)$, is an ensemble of $N(N-1)/2$ two-particle excitations. Each two-particle excitation, described by two-electron Hamiltonian, $\hat{H}_{il}\left(\vec{r}_i, \vec{r}_l\right)$, consists of two pseudo-electrons, in an effective pseudo-potential field, $\tilde{V}_{ext}\left(\vec{r}\right)$, plus the Coulomb interaction between these two pseudo-electrons. Since the indexes $\{i,\ l\}$ run from 1 to N, all Coulomb excitations are considered. Moreover, in a transformed lattice, *all* possible electron–electron Coulomb interactions reside solely *within* the two-particle excitations (**Figure 1**). Since there are no Coulomb interactions *between* the excitations, we can treat each two-particle excitation as a *quasi-closed subsystem,* with its two-particle wave function, $\phi_{il}\left(\vec{r}_i, \vec{r}_l\right)$. Consequently, in a transformed system, the two-particle representation is self-sufficient, with no room for a single-electron effective-field concept. Since all Coulomb interactions are included in this scheme, any additional corrections would be a double-counting mistake. Despite the lack of Coulomb interactions between the two-particle

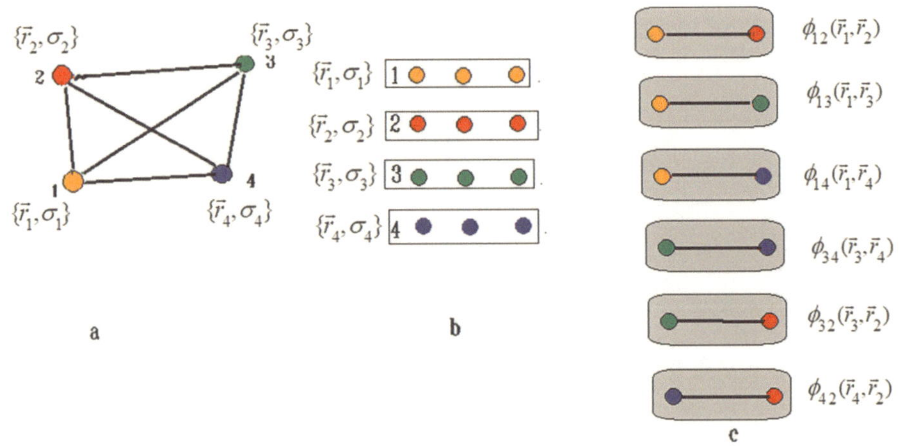

Figure 1.
Two-electron excitations in canonically transformed Hamiltonian (example: 4 electron system). (a) Original 4 electrons, with indexes $\{1,2,3,4\}$ *and space-spin coordinates* $\{\vec{r}_1,\sigma_1\}$, $\{\vec{r}_2,\sigma_2\}$, $\{\vec{r}_3,\sigma_3\}$, $\{\vec{r}_4,\sigma_4\}$. *Each electron has its own colour and 6 interactions between electrons are shown as black lines. (b) In canonically transformed Hamiltonian, each original electron splits into 3 pseudo-electrons which belong to the same group, and all pseudo-electrons have effective mass different from the effective mass of original electrons. All 3 electrons within each group, are described by the same spin-coordinates, group 1:* $\{\vec{r}_1,\sigma_1\}$, *group 2:* $\{\vec{r}_2,\sigma_2\}$, *group 3:* $\{\vec{r}_3,\sigma_3\}$, *group 4:* $\{\vec{r}_4,\sigma_4\}$. *(c) Within each group, there is no interactions between pseudo-electrons. Each pseudo-electron interacts solely with only one pseudo-electron from different group. There are 6 two-electron excitations. Two-electron excitation is a couple of pseudo-electrons from different group plus an interaction between pseudo-electrons (shown as black lines). The two-electron excitations are shown as 6 rounded rectangles, along with their two-electron wave functions,* $\phi_{12}(\vec{r}_1,\vec{r}_2), \phi_{13}(\vec{r}_1,\vec{r}_3), \phi_{14}(\vec{r}_1,\vec{r}_4), \phi_{34}(\vec{r}_3,\vec{r}_4), \phi_{32}(\vec{r}_3,\vec{r}_2), \phi_{42}(\vec{r}_4,\vec{r}_2)$.

excitations, they are still coupled by a different mechanism. Indeed, since each original electron, with an index S and the space-spin coordinate, $\{\vec{r}_s,\sigma_s\}$, is presented simultaneously in $(N-1)$ two-particle excitations, in a form of pseudo-electrons, the common space-spin coordinates couple excitations. This coupling forbids writing down the many-electron wave function, Ψ, as a product of two-electron wave functions, $\phi_{il}(\vec{r}_i,\vec{r}_l)$, and a different method is required to derive Ψ.

3. Fermionic many-electron wave function in terms of two-particle excitations

The unitary transformation, U, being applied to the Schrodinger equation,

$$\hat{H}\Psi = E\Psi \tag{7}$$

yields the transformed Schrodinger equation,

$$\tilde{\hat{H}}\tilde{\Psi} = \tilde{E}\tilde{\Psi} \tag{8}$$

$$\tilde{\Psi} = U\Psi U^{-1} \tag{9}$$

This unitary transformation has the following invariants: 1. The unitary transformation, $\tilde{E} = UEU^{-1} = E$, does not affect the energy of a system. This is an obvious invariant since re-grouping of the terms in $\hat{H}$ should not change the eigenvalues of $\hat{H}$. 2. The unitary transformed $\tilde{\Psi}_{\vec{k}_1\vec{k}_2.\dots\vec{k}_N}(\xi_1, \xi_2, \dots \xi_N)$ is a fermionic antisymmetric wave function, with respect to space-spin coordinates of a pseudo-electrons:

$$\tilde{\Psi}_{\vec{k}_1\vec{k}_2\dots\vec{k}_N}(\xi_1, \xi_2, ..\xi_i..\xi_l \dots \xi_N) = -\tilde{\Psi}_{\vec{k}_1\vec{k}_2\dots\vec{k}_N}(\xi_1, \xi_2, ..\xi_l..\xi_i..\xi_N) \tag{10}$$

This invariant follows from the definition of pseudo-electrons: the fermionic antisymmetry assumes that two pseudo-electrons of interest belong to different groups, i, l. 3. The unitary transformation does not affect the one-particle (free particle) wave functions, $\varphi^0_{s\vec{k}_s}(\xi_m) = \tilde{\varphi}^0_{s\vec{k}_s}(\xi_m)$, calculated, by definition, with no Coulomb interactions between electrons. Indeed, the transformation reduces both the kinetic energy operator and the potential energy operator by the same factor, $1/(N-1)$: $\hat{H}\varphi^0_{s\vec{k}_s} = E_n\varphi^0_{s\vec{k}_s}$; $\hat{H}/(N-1)\tilde{\varphi}^0_{s\vec{k}_s} = E_s/(N-1)\tilde{\varphi}^0_{s\vec{k}_s}$, and, hence,

$$\varphi^0_s(\xi_m) = \tilde{\varphi}^0_s(\xi_m) \tag{11}$$

4. The unitary transformation does not change the many-particle wave function, $\tilde{\Psi} = \Psi$, and expresses it in a different form. Indeed, since all $(N-1)$ pseudo-electrons, within the same group, have the same space-spin coordinates, the set, $\{\xi_1, \xi_2, \dots \xi_N\}$, of the coordinates of the electrons, coincides with the set of the coordinates of quasi-electrons, $\tilde{\Psi}_{\vec{k}_1\vec{k}_2.\dots\vec{k}_N}(\xi_1, \xi_2, \dots \xi_N)$. The QM defines the wave function as a probability, P, to find $i-th$ particle at coordinate ξ_i, within phase volume, $d\xi_i$ The original electron, with index i, has the same probability P with any of pseudo-electron within its group, because they have the same coordinate, ξ_i. Consequently, the unitary transformed wave function being expressed in terms of pseudo-electrons is an invariant: $\tilde{\Psi} = \Psi$. Since a many-particle system is an ensemble of two-particle excitations, $\Psi_{\vec{k}_1\vec{k}_2.\dots\vec{k}_N}(\xi_1, \xi_2, \dots \xi_N)$ can be expressed in terms of two-electron wave functions, $\phi_{i\vec{k}_i l\vec{k}_l}(\xi_k, \xi_m)$, $\{k, m\}$, $k < m$. One can prove from combinatorics of a Slater determinant that $\Psi_{\vec{k}_1\vec{k}_2.\dots\vec{k}_N}(\xi_1, \xi_2, \dots \xi_N)$ is a linear combination of two-particle wave functions, $\phi_{i\vec{k}_i l\vec{k}_l}(\xi_k, \xi_m)$, with coefficients, which are products of, $\varphi^0_{s\vec{k}_s}(\xi_s)$, by Levi-Civita symbols, $e^{i_1, i_2, \dots \dots i_N}$:

$$\Psi_{\vec{k}_1\vec{k}_2.\dots\vec{k}_N}(\xi_1, \xi_2, \dots \xi_N) = \sum_{\substack{\{i,l\} \\ i<l}} \sum_{\substack{\{k,m\} \\ k<m}} \mathrm{K}^k_i l^m \phi_{i\vec{k}_i l\vec{k}_l}(\xi_k, \xi_m)$$

$$\mathrm{K}^k_i l^m = \left(\frac{\sqrt{2}}{C^2_N\sqrt{N!}}\right) \times \sum_{\substack{i_1, i_2, ..i_N \\ no\, i_i \;\; and \;\; i_l}} e^{i_1, i_2, ..\mathrm{i}_i = \mathrm{k}, \dots \mathrm{i}_l = \mathrm{m}, \dots i_N} \prod_{\substack{s=1 \\ s\neq i \\ s\neq l}}^{N} \varphi^0_{s\vec{k}_s}(\xi_{i_s}))\} \tag{12}$$

Eq. (12) is an exact extension of a many-electron wave function in a *finite series* of two-particle wave functions. The derivation of Eq. (12) is based on a combinatorics of a Slater determinant with suppressed interactions, combined with the substitution $\phi_{\vec{ik}_i l\vec{k}_l}(\xi_k, \xi_m)$ instead of $\phi^0_{\vec{ik}_i l\vec{k}_l}(\xi_k, \xi_m)$.

4. Two-particle wave functions

The Hamiltonian of two pseudo-electrons in an effective potential, $\left(\tilde{V}_{ext}\left(\vec{r}_i\right) + \tilde{V}_{ext}\left(\vec{r}_l\right)\right)$, which interact by means of the Coulomb force is given by:

$$\hat{H}_{il}\left(\vec{r}_i, \vec{r}_l\right) = \sum_{i,l>i}\left(-\frac{\hbar^2}{2\tilde{m}}\nabla_i^2 - \frac{\hbar^2}{2\tilde{m}}\nabla_l^2 + \tilde{V}_{ext}\left(\vec{r}_i\right) + \tilde{V}_{ext}\left(\vec{r}_l\right) + \frac{e^2}{|\vec{r}_i - \vec{r}_l|}\right) \quad (13)$$

The two-electron wave functions, $\phi_{\vec{ik}_i l\vec{k}_l}\left(\vec{r}_i, \vec{r}_l\right)$, for either ortho or para spin configurations, are the solutions of the Schrodinger equation:

$$\hat{H}_{il}\left(\vec{r}_i, \vec{r}_l\right)\ \phi_{\vec{ik}_i l\vec{k}_l}\left(\vec{\mathrm{r}}_i, \vec{\mathrm{r}}_l\right) = E_{\vec{ik}_i l\vec{k}_k}\ \phi_{\vec{ik}_i l\vec{k}_l}\left(\vec{\mathrm{r}}_i, \vec{\mathrm{r}}_l\right) \quad (14)$$

It is known that the two-electron problem can be solved by different methods, see Ref. [22], and here two-particle Green function method is utilized. In this method, one needs first to suppress the Coulomb interaction between the electrons, or the last term in the right part of Eq. (13). As a result, the Hamiltonian becomes a sum of two one-particle Hamiltonians:

$$\begin{aligned}\hat{H}_{il}\left(\vec{r}_i, \vec{r}_l\right) &= \hat{H}_i\left(\vec{r}_i\right) + \hat{H}_l\left(\vec{r}_l\right),\\ \hat{H}_s\left(\vec{r}_s\right) &= -\frac{\hbar^2}{2\tilde{m}}\nabla_s^2 + \tilde{V}_{ext}\left(\vec{r}_s\right)\end{aligned} \quad (15)$$

The one-particle Green function, $G_E\left(\vec{r}, \vec{r}'\right)$, is built from the single-particle wave functions, $\varphi^0_{n\vec{k}}\left(\vec{r}\right)$:

$$G_E\left(\vec{r}, \vec{r}'\right) = -\sum_n\sum_{\vec{k}}\frac{\varphi^0_{n\vec{k}}\left(\vec{r}\right)\varphi^{0*}_{n\vec{k}}\left(\vec{r}'\right)}{\left(E - E_{n\vec{k}} - i\delta\right)} \quad (16)$$

where $\delta = 0^+$, and the sums are taken over all single-particle electronic states, $\left\{n, \vec{k}\right\}$ and all wave vectors, $\vec{k}$. The two-particle Green function, $G_E\left(\vec{r}_1, \vec{r}_2, \vec{r}'_1, \vec{r}'_2\right)$, is calculated from the one-particle Green function as an integral over the real axis, ε, of a complex plane energy plane, E:

$$G_E\left(\vec{r}_1, \vec{r}_2, \vec{r}'_1, \vec{r}'_2\right) = -\frac{1}{2\pi i}\int_{-\infty}^{+\infty} d\varepsilon\, G_E\left(\vec{r}_1, \vec{r}'_1\right) G_{E-\varepsilon}\left(\vec{r}_2, \vec{r}'_2\right) \quad (17)$$

The calculation of an integral in the right part of Eq. (17), with the help of Eq. (16), provides the following result:

$$G_E\left(\vec{r}_1, \vec{r}_1{}'; \vec{r}_2, \vec{r}_2{}'\right) = 2\sum_s \sum_p \sum_{\vec{k}_s} \sum_{\vec{k}_p} \frac{\varphi_{s\vec{k}_s 0}\left(\vec{r}_1\right) \varphi^{0*}_{s\vec{k}_s}\left(\vec{r}_1{}'\right) \varphi_{p\vec{k}_p 0}\left(\vec{r}_2\right) \varphi_{p\vec{k}_p 0*}\left(\vec{r}_2{}'\right)}{\left(E - E_{s\vec{k}_s} - E_{p\vec{k}_p} - i\delta\right)} \tag{18}$$

Next, in an accordance with the Green function method, in Eq. (13), we should turn on the Coulomb interactions, $H^{il}_{\text{int}} = e^2/|\vec{r}_i - \vec{r}_l|$, and calculate the corrected two-electron wave functions, $\phi_{il}\left(\vec{r}_i, \vec{r}_l\right)$. In symbol, $\phi_{il}\left(\vec{r}_i, \vec{r}_l\right)$, for simplicity, we have omitted the $\vec{k}$ vector and $\phi_{il}\left(\vec{r}_i, \vec{r}_l\right)$ read as $\phi_{i\vec{k}_i l\vec{k}_l}\left(\vec{r}_i, \vec{r}_l\right)$. Based on the Green function method, the corrected two-electron wave functions, $\phi_{il}\left(\vec{r}_i, \vec{r}_l\right)$, satisfy the following integral equation:

$$\phi_{il}\left(\vec{r}_i, \vec{r}_l\right) = \phi^0{}_{il}\left(\vec{r}_i, \vec{r}_l\right) + \Delta\phi_{il}\left(\vec{r}_i, \vec{r}_l\right) \tag{19}$$

$$\begin{gathered}\phi_{il}\left(\vec{r}_i, \vec{r}_l\right) = \phi^0{}_{il}\left(\vec{r}_i, \vec{r}_l\right) - \iint d\vec{r}'_i d\vec{r}'_l G_E\left(\vec{r}_1, \vec{r}_1{}'; \vec{r}_2, \vec{r}_2{}'\right) \mathrm{H}^{il}_{\text{int}}\left(\vec{r}'_i, \vec{r}'_l\right) \phi_{il}\left(\vec{r}'_i, \vec{r}'_l\right), \\ E = E_{i\vec{k}_i} + E_{l\vec{k}_l}; \{sp\} \neq \{il\}, \\ \{ps\} \neq \{il\}\end{gathered} \tag{20}$$

Hence, the corrections, $\Delta\phi_{il}\left(\vec{r}_i, \vec{r}_l\right)$, to undisturbed two-particle wave functions obey the equation:

$$\Delta\phi_{il}\left(\vec{r}_i, \vec{r}_l\right) = -\iint d\vec{r}'_i d\vec{r}'_l G_E\left(\vec{r}_1, \vec{r}_1{}'; \vec{r}_2, \vec{r}_2{}'\right) \mathrm{H}^{il}_{\text{int}}\left(\vec{r}'_i, \vec{r}'_l\right) \left[\phi^0{}_{il}\left(\vec{r}'_i, \vec{r}'_l\right) + \Delta\phi_{il}\left(\vec{r}'_i, \vec{r}'_l\right)\right] \tag{21}$$

The integral Eq. (21) can be solved by different methods, e.c. by the iteration method:

$$\begin{gathered}\Delta\phi^{(1)}{}_{il}\left(\vec{r}_i, \vec{r}_l\right) = -\iint d\vec{r}'_i d\vec{r}'_l G_E\left(\vec{r}_1, \vec{r}_1{}'; \vec{r}_2, \vec{r}_2{}'\right) \mathrm{H}^{il}_{\text{int}}\left(\vec{r}'_i, \vec{r}'_l\right) \phi^0{}_{il}\left(\vec{r}'_i, \vec{r}'_l\right) \\ \Delta\phi^{(2)}{}_{il}\left(\vec{r}_i, \vec{r}_l\right) = -\iint d\vec{r}'_i d\vec{r}'_l G_E\left(\vec{r}_1, \vec{r}_1{}'; \vec{r}_2, \vec{r}_2{}'\right) \mathrm{H}^{il}_{\text{int}}\left(\vec{r}'_i, \vec{r}'_l\right) \left[\phi^0{}_{il}\left(\vec{r}'_i, \vec{r}'_l\right) + \Delta\phi^{(1)}{}_{il}\left(\vec{r}'_i, \vec{r}'_l\right)\right] \\ \ldots \ldots \ldots \ldots \ldots \ldots \ldots \ldots \ldots \ldots \ldots \\ \Delta\phi^{(n)}{}_{il}\left(\vec{r}_i, \vec{r}_l\right) = -\iint d\vec{r}'_i d\vec{r}'_l G_E\left(\vec{r}_1, \vec{r}_1{}'; \vec{r}_2, \vec{r}_2{}'\right) \mathrm{H}^{il}_{\text{int}}\left(\vec{r}'_i, \vec{r}'_l\right) \left[\phi^0{}_{il}\left(\vec{r}'_i, \vec{r}'_l\right) + \Delta\phi^{(n-1)}{}_{il}\left(\vec{r}'_i, \vec{r}'_l\right)\right]\end{gathered} \tag{22}$$

This procedure, after n iterations, yields $\Delta\phi_{il}\left(\vec{r}_i, \vec{r}_l\right) \approx \Delta\phi^{(n)}{}_{il}\left(\vec{r}_i, \vec{r}_l\right)$, a desirable accuracy of $(n-1)$ order with respect to the Coulomb interaction, H^{il}_{int}.

5. Energy functional and quasi-particle energies

The quasi-particle energy, $E_i^{QP}\left(\vec{k}_i\right)$, is either the energy of the electron, $P = E$, or the energy of the hole, $P = H$, in $i - th$ state, with the wave vector, $\vec{k}_i$. $E_i^{QP}\left(\vec{k}_i\right)$, can be obtained from the expression for the electronic part of the energy of a lattice, E_T (energy functional):

$$\mathrm{E}_T = NORM * \iiint \tilde{\Psi}^* (\xi_1, \ldots \xi_N) \tilde{\hat{\mathrm{H}}} \tilde{\Psi}(\xi_1, \ldots \xi_N) d\xi_1 d\xi_2 \ldots d\xi_N, \qquad NORM = \frac{\Delta V_0^N}{N(2\pi)^{3N}} \tag{23}$$

Here $NORM$ is the normalization, and ΔV_0 is the volume of the elementary cell. The substitution of Eq. (12) into Eq. (23) yields E_T as a sum of five contributions:

$$E_T = NORM * \left(E_0 + E_{e-e} + E_0^{Scr} + E_{e-e}^{Scr} + E_{SE} \right) \tag{24}$$

The contributions to the E_T are associated with the following interactions: E_0 is the total energy of the electrons in an effective potential, with the suppressed Coulomb interactions between electrons. E_{e-e} is the energy of the Coulomb interactions between electrons, E_{e-e}, with no screening. E_0^{Scr} is the correction to the E_0 due to the screening. E_{e-e}^{Scr} is the correction to the E_{e-e} due to the screening. E_{SE} is the exchange-correlation energy, or the so-called self-energy, Σ_{SE}. The expressions for these contributions are mentioned below:

$$E_0 = \iint \ldots \int \Psi^{0*}(\xi_1, ..\xi_N) \sum_s \hat{H}_s \, \Psi^0(\xi_1, ..\xi_N) \mathrm{d}\xi_1, ..d\xi_N \tag{25}$$

$$E_{e-e} = \iint \ldots \int \Psi^{0*}(\xi_1, ..\xi_N) \sum_{\substack{p,q \\ p<q}} H_{\text{int}}^{pq} \, \Psi^0(\xi_1, ..\xi_N) \mathrm{d}\xi_1, ..d\xi_N \tag{26}$$

$$\begin{aligned} E_0^{Scr} = &\iint \ldots \int \Psi^{0*}(\xi_1, ..\xi_N) \sum_s \hat{H}_s \sum_{\substack{i,l \\ i<l}} \sum_{\substack{k,m \\ k<m}} K_{il}^{km} \Delta\phi_{il}(\xi_k, \xi_m) \mathrm{d}\xi_1, ..d\xi_N + \\ &+ \iint \ldots \int \sum_{\substack{i,l \\ i<l}} \sum_{\substack{k,m \\ k<m}} K_{il}^{km} \Delta\phi_{il}(\xi_k, \xi_m) \sum_s \hat{H}_s \Psi^{0*}(\xi_1, ..\xi_N) \quad \mathrm{d}\xi_1, ..d\xi_N \end{aligned} \tag{27}$$

$$\begin{aligned} E_{e-e}^{Scr} = &\iint \ldots \int \Psi^{0*}(\xi_1, ..\xi_N) \sum_{\substack{p,q \\ p<q}} H_{\text{int}}^{pq} \sum_{\substack{\mathrm{i,l} \\ i<l}} \sum_{\substack{\mathrm{k,m} \\ k<m}} K_{il}^{km} \Delta\phi_{il}(\xi_k, \xi_m) \mathrm{d}\xi_1, ..d\xi_N + \\ &+ \iint \ldots \int \sum_{\substack{\mathrm{i,l} \\ i<l}} \sum_{\substack{\mathrm{k,m} \\ k<m}} K_{il}^{km} \Delta\phi_{il}(\xi_k, \xi_m) \sum_{\substack{p,q \\ p<q}} H_{\text{int}}^{pq} \Psi^{0*}(\xi_1, ..\xi_N) \quad \mathrm{d}\xi_1, ..d\xi_N \end{aligned} \tag{28}$$

$$E_{SE} = \iint \cdot \int \sum_{\substack{i,l \\ i<l}} \sum_{\substack{k,m \\ k<m}} K_{il}^{*km} \Delta\phi_{il}^{*}(\xi_k, \xi_m) \sum_s \hat{H}_s \sum_{\substack{i,l \\ i<l}} \sum_{\substack{k,m \\ k<m}} K_{il}^{km} \Delta\phi_{il}(\xi_k, \xi_m) \mathrm{d}\xi_1, .d\xi_N +$$
$$+ \iint \cdot \int \sum_{\substack{i,l \\ i<l}} \sum_{\substack{k,m \\ k<m}} K_{il}^{*km} \Delta\phi_{il}^{*}(\xi_k, \xi_m) \sum_{\substack{p,q \\ p<q}} H_{\text{int}}^{pq} \sum_{\substack{i,l \\ i<l}} \sum_{\substack{k,m \\ k<m}} K_{il}^{km} \Delta\phi_{il}(\xi_k, \xi_m) \mathrm{d}\xi_1, .d\xi_N \tag{29}$$

The expressions (25)–(29) are reduced, after considerable analytical efforts, to the closed integral forms:

$$E_0 = \sum_i \sum_{\vec{k}_i} E_i\left(\vec{k}_i\right) \tag{30}$$

$$E_{e-e} = \frac{2e^2(N-2)!}{N!} \sum_{\vec{k}_i} \sum_{\vec{k}_l} \sum_{i,l>i} \sum_{r,t>r} \iint \frac{\left|\phi^0_{i\vec{k}_i l\vec{k}_l}\left(\vec{\mathrm{r}}_r, \vec{\mathrm{r}}_t\right)\right|^2}{\left|\vec{r}_r - \vec{r}_t\right|} \, \mathrm{d}\vec{\mathrm{r}}_r d\vec{r}_t \tag{31}$$

$$E_0^{Scr} = \frac{4E_0}{N^2(N-1)^2} \sum_{\vec{k}_i} \sum_{\vec{k}_l} \sum_{i,l>i} \sum_{r,t>r} \iint \Delta\phi^{*}_{i\vec{k}_i l\vec{k}_l}\left(\vec{\mathrm{r}}_r, \vec{\mathrm{r}}_t\right) \phi^{0}_{i\vec{k}_i l\vec{k}_l}\left(\vec{\mathrm{r}}_r, \vec{\mathrm{r}}_t\right) \mathrm{d}\vec{\mathrm{r}}_r d\vec{r}_t + C.C. \tag{32}$$

$$E_{e-e}^{SCR} = \frac{4e^2(N-4)!}{N!C_N^2} \sum_{\vec{k}_i} \sum_{\vec{k}_l} \sum_{i,l>i} \sum_{r,t>r} \left[\iint \Delta\phi^{*}_{i\vec{k}_i l\vec{k}_l}\left(\vec{\mathrm{r}}_r, \vec{\mathrm{r}}_t\right) \phi^{0}_{i\vec{k}_i l\vec{k}_l}\left(\vec{\mathrm{r}}_r, \vec{\mathrm{r}}_t\right) \mathrm{d}\vec{\mathrm{r}}_r d\vec{r}_t \right.$$
$$\left. \times \sum_{\vec{k}_s} \sum_{\vec{k}_{s'}} \left(\sum_{\substack{s,s' \\ s>s' \\ \{SS'\}\neq\{il\}}} \sum_{\substack{p,q \\ p>q,\ \{pq\}\neq\{rt\}}} \iint \frac{\left|\phi^0_{s\vec{k}_s s'\vec{k}_{s'}}\left(\vec{\mathrm{r}}_p, \vec{\mathrm{r}}_q\right)\right|^2}{\left|\vec{\mathrm{r}}_p - \vec{r}_q\right|} \mathrm{d}\vec{\mathrm{r}}_p d\vec{r}_q \right) \right]$$
$$+ \frac{2e^2(N-2)!}{N!C_N^2} \sum_{\vec{k}_i} \sum_{\vec{k}_l} \sum_{i,l>i} \sum_{r,t>r} \iint \frac{\Delta\phi^{*}_{i\vec{k}_i l\vec{k}_l}\left(\vec{\mathrm{r}}_r, \vec{\mathrm{r}}_t\right) \phi^{0}_{i\vec{k}_i l\vec{k}_l}\left(\vec{\mathrm{r}}_r, \vec{\mathrm{r}}_t\right)}{\left|\vec{\mathrm{r}}_r - \vec{r}_t\right|} \mathrm{d}\vec{\mathrm{r}}_r d\vec{r}_t + C.C. \tag{33}$$

$$\begin{aligned} E_{SE} &= E_{SE}(i,i';l,l';r,r';t,t' + E_{SE}(i,i;l,l;r,r;t,t) + \\ &+ E_{SE}(i,i;l,l;r,r';t,t') + E_{SE}(i,i;l,l;r,r;t,t') + \\ &+ E_{SE}\left(i,i;l,l';r,r;t,t'\right) \end{aligned} \tag{34}$$

where:

$$E_{SE}(i,i;l,l;r,r;t,t) = \sum_{i,l>i;}\sum_{\vec{k}_i}\sum_{\vec{k}_l}\sum_{r,t>r}\left\{\left[E_0 - \frac{(N-3)}{(N-2)}\left(E^0_{i\vec{k}_i} + E^0_{l\vec{k}_l}\right)\right]\right.$$

$$\times \frac{8(N-2)}{N^3(N-1)^3}\iint\left|\Delta\phi_{i\vec{k}_il\vec{k}_l}\left(\vec{r}_r,\vec{r}_t\right)\right|^2 d\vec{r}_r d\vec{r}_t \tag{35}$$

$$\left.-\frac{8e^2}{N^3(N-1)^3}\iint\frac{\Delta\phi_{i\vec{k}_il\vec{k}_l}\left(\vec{r}_r,\vec{r}_t\right)\phi^{0\,*}_{i\vec{k}_il\vec{k}_l}\left(\vec{r}_r,\vec{r}_t\right)d\vec{r}_r d\vec{r}_t}{\left|\vec{r}_r-\vec{r}_t\right|}\right\}$$

$$E_{SE}\left(i,i';l,l';r,r';t,t'\right) = \sum_{i,l>i;}\ \sum_{\substack{i'\neq i;\\ l'\neq l;\\ l'>i'}}\ \sum_{r,t>r}\ \sum_{\substack{r'\neq r,t'\neq t\\ t'>r'}}\ \sum_{\vec{k}_i}\sum_{\vec{k}_l}\sum_{\vec{k}_{i'}}\sum_{\vec{k}_{l'}}$$

$$\times\left\{\frac{8}{N^3(N-1)^3(N-2)}\left[E_0 - \frac{(N-5)}{(N-3)}\left(E^0_{i'\vec{k}_{i'}} + E^0_{l'\vec{k}_{l'}}\right)\right]\times\right.$$

$$\times\left[\iint\Delta\phi_{i\vec{k}_i l\vec{k}_l}\left(\vec{r}_r,\vec{r}_t\right)\phi^{0\,*}_{i\vec{k}_i l\vec{k}_l}\left(\vec{r}_r,\vec{r}_t\right)d\vec{r}_r d\vec{r}_t\right]\times \tag{36}$$

$$\left[\iint\Delta\phi^*_{i'\vec{k}'_i l'\vec{k}_{l'}}\left(\vec{r}_{r'},\vec{r}_{t'}\right)\phi^0_{i'\vec{k}_{i'}l'\vec{k}_{l'}}\left(\vec{r}_{r'},\vec{r}_{t'}\right)d\vec{r}_{r'}d\vec{r}_{t'}\right] -$$

$$-\left(\frac{16e^2}{N^3(N-1)^3(N-2)(N-3)}\right)\left[\iint\frac{|\phi^0_{i'\vec{k}_{i'}l'\vec{k}_{l'}}\left(\vec{r}_{r'},\vec{r}_{t'}\right)|^2 d\vec{r}_{r'}\vec{r}_{t'}}{|\vec{r}_{r'}-\vec{r}_{t'}|}\right]\times$$

$$\left.\times\left[\iint\Delta\phi_{i\vec{k}_i l\vec{k}_l}\left(\vec{r}_r,\vec{r}_t\right)\phi^{0\,*}_{i\vec{k}_i l\vec{k}_l}\left(\vec{r}_r,\vec{r}_t\right)d\vec{r}_r d\vec{r}_t\right]\right\}$$

$$E_{SE}(i,i;l,l;r,r;t,t') = \sum_{i,l>i;}\ \sum_{r,t>r}\ \sum_{t'\neq t}\sum_{\vec{k}_i}\sum_{\vec{k}_l}\sum_{\vec{k}_p}\sum_{\substack{p\neq i\\ p\neq l}}\times\left\{\frac{-8}{N^3(N-1)^3(N-2)}\right.$$

$$\times\left(E_0 - \frac{(N-2)}{(N-1)}\left(E^0_{i\vec{k}_i} + E^0_{l\vec{k}_l}\right)\right)\left[\iint\Delta\phi_{i\vec{k}_il\vec{k}_l}\left(\vec{r}_r,\vec{r}_t\right)\varphi^{0*}_{p\vec{k}_p}\left(\vec{r}_r,\vec{r}_t\right)d\vec{r}_r d\vec{r}_t\right]\times$$

$$\times\left[\iint\Delta\phi^*_{i\vec{k}_il\vec{k}_l}\left(\vec{r}_r,\vec{r}_{t'}\right)\varphi^0_{p\vec{k}_p}\left(\vec{r}_r,\vec{r}_{t'}\right)d\vec{r}_r d\vec{r}_{t'}\right]+ \tag{37}$$

$$+\frac{8e^2}{N^3(N-1)^3(N-2)}\int d\vec{r}_r\left(\left[\int\left(\Delta\phi_{i\vec{k}_il\vec{k}_l}\left(\vec{r}_r,\vec{r}_t\right)\varphi^{0*}_{p\vec{k}_p}\left(\vec{r}_r,\vec{r}_t\right)d\vec{r}_t\right]\times\right.$$

$$\left.\left.\left[\int\frac{\Delta\phi^*_{i\vec{k}_il\vec{k}_l}\left(\vec{r}_r,\vec{r}_t\right)\varphi^0_{p\vec{k}_p}\left(\vec{r}_r,\vec{r}_t\right)d\vec{r}_t}{|\vec{r}_r-\vec{r}_t|}\right]\right)\right\}$$

$$E_{SE}(i,i;l,l;r,r';t,t') = \sum_{i,l>i;} \sum_{r,t>r} \sum_{\substack{r'\neq r,t'\neq t\\ t'>r'}} \sum_{\substack{q,p''>q\\ \{qp''\}\neq\{il\}}} \sum_{\vec{k}_i} \sum_{\vec{k}_l} \sum_{\vec{k}_{p''}} \sum_{\vec{k}_q} \left\{ \frac{8/(N-1)^3}{N^3(N-2)} \right.$$

$$\times \left[E_0 + \frac{(N-4)}{2} \left(E^0{}_{ik_i} + E^0{}_{lk_l}\right) \right] \iint \Delta\phi_{i\vec{k}_i l\vec{k}_l}\left(\vec{r}_r, \vec{r}_t\right) \phi^{0*}_{q\vec{k}_q p''\vec{k}_{p''}}\left(\vec{r}_r, \vec{r}_t\right) d\vec{r}_r d\vec{r}_t$$

$$\times \left[\iint \Delta\phi^*{}_{i\vec{k}_i l\vec{k}_l}\left(\vec{r}_{r'}, \vec{r}_{t'}\right) \phi^0_{q\vec{k}_q p''\vec{k}_{p''}}\left(\vec{r}_{r'}, \vec{r}_{t'}\right) d\vec{r}_{r'} d\vec{r}_{t'} \right] - \frac{8e^2}{N^3(N-1)^3}$$

$$\left. \times \iint \Delta\phi_{i\vec{k}_i l\vec{k}_l}\left(\vec{r}_r, \vec{r}_t\right) \phi^{0*}_{q\vec{k}_q p''\vec{k}_{p''}}\left(\vec{r}_r, \vec{r}_t\right) d\vec{r}_r d\vec{r}_t \iint \phi^*{}_{i\vec{k}_i l\vec{k}_l}\left(\vec{r}_{r'}, \vec{r}_{t'}\right) \frac{\phi^0_{q\vec{k}_q p''\vec{k}_{p''}}\left(\vec{r}_{r'}, \vec{r}_{t'}\right)}{\left|\vec{r}_{r'} - \vec{r}_{t'}\right|} d\vec{r}_{r'} d\vec{r}_{t'} \right\} \tag{38}$$

$$E_{SE}\left(i,i;l,l';r,r;t,t'\right) = \sum_{i,l>i;} \sum_{\substack{l'>i';\\ i'\neq i\\ l'\neq l}} \sum_{r,t>r} \sum_{\substack{t'\neq t\\ t'>r}} \sum_{\vec{k}_i} \sum_{\vec{k}_l} \sum_{\vec{k}_{l'}} \left\{ \frac{8(N-4)}{N^3(N-1)^3(N-2)} \times \right.$$

$$\times \left[2E_0 - \left(E^0_{i\vec{k}_i} + E^0_{l\vec{k}_l} \right) \right] \times \int d\vec{r}_r \left\{ \left[\int \varphi^{0*}_{l\vec{k}_l}\left(\vec{r}_t\right) \Delta\phi_{i\vec{k}_i l\vec{k}_l}\left(\vec{r}_r, \vec{r}_t\right) d\vec{r}_t \right] \times \right.$$

$$\left. \int d\vec{r}_{t'} \left[\int \varphi^0_{l'\vec{k}_{l'}}\left(\vec{r}_{t'}\right) \Delta\phi^*{}_{i\vec{k}_i l'\vec{k}_{l'}}\left(\vec{r}_r, \vec{r}_{t'}\right) d\vec{r}_{t'} \right] \right\} - \frac{8e^2}{N^3(N-1)^3(N-2)}$$

$$\left. - \int d\vec{r}_r \left\{ \left[\int \varphi^{0*}_{l\vec{k}_l}\left(\vec{r}_t\right) \Delta\phi_{i\vec{k}_i l\vec{k}_l}\left(\vec{r}_r, \vec{r}_t\right) d\vec{r}_t \right] \int d\vec{r}_{t'} \left[\int \frac{\varphi^0_{l'\vec{k}_{l'}}\left(\vec{r}_{t'}\right) \phi^*{}_{i\vec{k}_i l'\vec{k}_{l'}}\left(\vec{r}_r, \vec{r}_{t'}\right) d\vec{r}_{t'}}{\left|\vec{r}_r - \vec{r}_{t'}\right|} \right] \right\} \right\} \tag{39}$$

As we mentioned above, the quasi-particle energies, $E_i^{QP}\left(\vec{k}_i\right)$, can be obtained from Eqs. (30)–(39) as a sum of terms of their right parts with a fixed index i.

6. Discussion and conclusions

The many-electron problem has a unique exact analytical solution [Eq. (12)], or many-electron wave function, Ψ, which is a linear combination, with a finite number of terms, of two-particle wave functions. This result stands for *any magnitude* of interactions between particles, and no higher-order excitations are required to be included in the scheme. Moreover, one can see from Eq. (4) that, in a unitary transformed lattice, the electronic excitations with an order larger than two simply do not exist. Generally, a developed scheme is a natural consequence of the fact that any elementary interaction includes two and only two particles. It is worth remembering that the two-particle wave functions, ϕ_{il}, in Eq. (12), describe the pseudo-electrons in a unitary transformed lattice, and *not* the original electrons. This point is an essential novelty of our theory. Indeed, two-electron wave functions are presented in other many-body theories [5–8, 10–13], but these representations always assume truncations or neglect the terms in a Hamiltonian. In an analytical part of our calculations, the tremendous help comes from the specific form of Eq. (12) for the many-body wave function, Ψ. In Eq. (12), the coefficients, K_{il}^{km}, in front of the two-particle wave function, ϕ_{il}, are the products of the ortho-normalized non-disturbed (free) one-particle wave functions, $\varphi^0_{s\vec{k}_s}$. Consequently, in the calculation of a physical measurable, $P = \int \Psi^* \hat{P}\Psi dq$ (energy functional), many integrals of products, $\varphi^0_{s\vec{k}_s}\varphi^{0*}_{s\vec{k}_{s'}}$ are zeros. As a result, the analytical expressions for the integrals can be relatively easily derived. In this chapter, just as an example, we choose a physical measurable P as an energy of a crystal, and express this energy in a closed form, in terms of 6D space-coordinates integrals. Since the wave function, Ψ, is known, a similar procedure can be carried out for any physical measurable of interest, such as momentum, impulse, currents, and conductivity ...

It would be instructive to compare our theory with the DFT scheme. First, the original DFT method was developed for ground states only, and our approach works universally for *any state*, no matter ground or excited. The Eqs. (30)–(31), for both the total energy, E_0, of the electrons with the suppressed Coulomb interactions between the electrons, and for the energy of the Coulomb interactions between electrons, E_{e-e}, with no screening, are identical to the calculation of DFT scheme [5–8]. Since our result stands for all states, no matter ground or excited, the DFT theory exactly describes the energy, E_{e-e}, for both *ground and excited states*, the result we could not find in literature. The differences between DFT and our approach are laid within the most delicate parts of the energy functional, i.e., the screening correction, E_0^{Scr}, [Eq. (32)], the screening correction, E_{e-e}^{Scr}, to the E_{e-e}, [Eq. (33)], and the exchange-correlation energy, E_{SE} , or the so-called self-energy, Σ_{SE}, [Eqs. (34)–(39)]. It is known that the calculation of the exchange-correlation energy is the most difficult problem in first-principles calculations [23]. In our scheme, the exchange-correlation energy appears as a unique closed integral form, without any additional approximations and efforts, which are required, for example, in Hedin's calculations for the exchange-correlation energy [19]. Moreover, in a proposed scheme, in contrast to DFT and Hedin GW [19], both ground state and the excited states are treated on the same footing, within the same QM scheme. Additionally, since in our theory, all Coulomb interactions are treated on the same footing, we do not separate the "direct" Coulomb interactions from the "screened" Coulomb interactions. Still, in Eqs. (24)–(29) for the total crystal energy, the screened, the direct, and the exchange-correlation interactions appear as profoundly separate contributions. There are two additional advantages of our approach in comparison with the Density Functional schemes [5–9] and other methods based on an effective-field concept [10–13]. In these schemes, the band structure and other physical properties are calculated by adding or removing an electron, or a hole to the system. This procedure inevitably causes the problem of a *response* of a many-body system, due to the relaxation of the atomic orbitals and the screening. Note, that the problem of a response is *entirely* due to one-particle-in-effective-field concept utilized in these theories. This response is usually described by means of the dielectric matrix, $\hat{\varepsilon}$, which is to be calculated self-consistently [3]. Consequently, the singe-particle nature of these schemes brings additional difficulties in defining the consistent quasi-macroscopic dielectric function. Since in our method we do not rely on the single-particle effective-field model, the problem of response simply does not exist, with no need for the dielectric relaxation matrix, $\hat{\varepsilon}$. Moreover, in a suggested method, the many-electron wave function provides a self-sufficient input for calculation of any measurable, with no need for inclusion of hole–electron interactions which would be a double-counting mistake.

Author details

Adil-Gerai Kussow
Department of Physics, University of Connecticut, Storrs, CT, USA

*Address all correspondence to: akussow@yahoo.com; adil_gerai.kussow@uconn.edu

References

[1] Dirac PAM. Mint: Quantum mechanics of many-electron systems. Proceedings of the Royal Society of London. Series A. 1929;**123**:714-715. DOI: 10.1098/rspa./929.0094

[2] Lev K. Quantum Theory of the Solid State: An Introduction. Dordrecht: Springer Science +Business Media, B.V; 2004. p. 282. DOI: 10.1007/978-1-4020-2154-1

[3] Deslippe J, Samsonidze G., Strub D. A, Jain M, Cohen M.L, and Louie S.G: Berkley GW: A massive parallel computer package for the calculation of the quasiparticle and optical properties of materials and nanostructures, Computer Physics Communication, 2012;**183**:1269-1289. DOI: 10.1016/j.cpc.2011.12.006

[4] Silverstone HJ, Sinanoglu O. Many-electron theory of nonclosed-shell atoms and molecules. I. Orbital wave functions and perturbation theory. The Journal of Chemical Physics. 1966;**44**:1899-2002. DOI: 10.1063/1.1726959

[5] Kohn W. Nobel lecture: Electronic structure of matter-wave functions and density functionals. Reviews of Modern Physics. 1999;**71**:1253-1266. DOI: 10.1103/RevModPhys.71.1253

[6] Kohn W, Sham LJ. Self-consistent equations including exchange and correlation effects. Physical Review A. 1965;**140**:1133-1138. DOI: 10.1103/Phys.Rev.140.A1133

[7] Dreizler RM, Gross RM. Density Functional Theory. New York: Springer Science + Business Media; 1995. p. 671. DOI: 10:1007/978-1-4757-99750

[8] Petersilka M, Gossmann UJ, Gross EKU. Excitation energies from time-dependent density-functional theory. Physical Review Letters. 1996;**76**: 1212-1215. DOI: 10.1103/PhysRevLett.76.1212

[9] Lundqvist S, March NH. Theory of Inhomogeneous Electron Gas. New York: Springer Science +Business Media; 1983. p. 303. DOI: 10.1007/978-1-4899-0415-7

[10] Bishop RF. An overview of coupled cluster theory and its applications in physics. Theoretical Chemistry Accounts. 1991;**80**:95-148. DOI: 10.1007/BF01119617

[11] Coester F. Bound states f a many-particle system. Nuclear Physics. 1958; **47**:421-424. DOI: 10.1016/0029-5582(58)90280-2

[12] Coester F, Kummel H. Short-range correlations in nuclear wave functions. Nuclear Physics. 1960;**17**:477-485. DOI: 10.1016./0029-5582(60)90140-I

[13] Cizek J. On the correlation problem in atomic and molecular systems. Calculation of wavefunctions components in Ursell-type expansion using Quantum-Field theoretical methods. The Journal of Chemical Physics. 1966;**45**:4256-4264. DOI: 10.1063/1.1727484

[14] Bethe HA, Salpeter EE. Quantum Mechanics of One and Two-Electron Atoms. New York: Plenum Publishing Corporation; 1977. p. 370. DOI: 10.1007/978-1-4613-4104-8

[15] Surjan P R. Correlation and localization. Berlin-Heidelerg: Springer-Verlag; 1999. p. 325. DOI: 10.1007/3-540-48072-X_4

[16] Werner H J, Knizia G, Manby F R: Explicitly correlated cluster methods

with pair-specific geminals. Molecular Physics. 2011;**109**:407-417. DOI: doi.org/10.1080/00268976.2010.526641

[17] Dirac PAM. On the nature of quantum mechanics. Proceedings of Royal Society. 1926;**112**:762-775. DOI: 10.1098/rspa.1926.0133

[18] Dirac PAM. On the analogy between classical and quantum mechanics. Reviews of Modern Physics. 1945;**17**: 195-220. DOI: 10/1103/RevModPhys.17.195

[19] Hedin L. New method for calculating the one-particle Green function with application to the electron gas problem. Physical Review A. 1965;**139**:796-812. DOI: 10.1103/PhysRev.139.A796

[20] Rohlfing M, Louie SG. Electron-hole excitations and optical spectra from first principles. Physical Review B. 2000;**62**: 4927-4944. DOI: 10.1103/PhysRevB.62.492

[21] Rohlfing M, Louie SG. Electron-hole excitations in semiconductors and insulators. Physical Review Letters. 1988;**81**:2312-2315

[22] Bethe HA, Salpeter EE. Quantum Mechanics of One-and Two-Electron Atoms. New York: A Plenum Rosetta Edition; 1977. p. 370. DOI: 10.1007/1978-1-4613-4104-8

[23] Onida G, Reining L, Rubio A. Electronic excitations: density functional versus many-body Green-function approach. Reviews of Modern Physics. 2002;**74**:601-657. DOI: 10.1103/RevModPhys.74.601

Chapter 2

SOA Model and Design Guidelines in Lossless Photonic Subsystem

Pantea Nadimi Goki, Antonio Tufano, Fabio Cavaliere and Luca Potì

Abstract

We propose a new practical analytical model to calculate the performance of amplitude-modulated systems, including semiconductor optical amplifiers (SOA). Lower and upper-performance bounds are given in terms of signal quality factor (Q) concerning the input signal pattern. The target is to provide a design tool for gain elements included in photonic integrated circuits (PIC) to compensate for their insertion loss. This subject is a critical issue, for example, in the arrays of optical transmitters with silicon photonics modulators used for interconnection applications. Due to implementation limitations, the design of an SOA embedded in a PIC is considerably different with respect to the use of SOAs as line amplifiers in optical networks. SOA amplified spontaneous emission (ASE) and gain saturation effects have been included in the model, together with the input signal extinction ratio and the receiver electrical filter. Each degradation effect provides its own contribution to the signal integrity in terms of signal-to-noise ratio (SNR) or inter-symbol interference (ISI). The model shows that the SOA operation at low extinction ratios, typical in optical interconnect applications, is substantially different from the operation at higher extinction ratios used in transport networks. The model is validated through numerical simulations and experiments. Finally, two examples are provided for dimensioning a PIC system and optimizing the SOA parameters.

Keywords: extinction ratio, filtering, probability density function, Q-factor, semiconductor optical amplifiers

1. Introduction

The growing demand for fast, miniaturized, and power-efficient optical devices leads to provide growth opportunities for photonic integrated circuits (PICs) to process or distribute information. Passive PICs are widely utilized, for example, for optical beam steering [1], integrated photonic filters [2], and integrated silicon photonics transceiver [3] nevertheless, active PICs have numerous applications in optical systems such as lossless reconfigurable optical add-drop multiplexers (ROADM) in the fronthaul network [4]. The PICs allow optical systems to be more compact than discrete optical components. However, some PIC components, such as modulators, multiplexers, and splitters, cause a high loss. In addition, in a PIC, active and passive optical components are interconnected by lossy optical waveguides and need to be coupled with input and output fibers, which adds even more attenuation. One of the most promising solutions for compensating PIC losses is to integrate a semiconductor optical amplifier (SOA) on-chip in combination with other passive components. Numerous research studies have been reported on SOA technology and applications, both as on-chip or stand-alone components [5, 6]. SOAs can be designed to amplify signals in a very wide band range, over 100 nm optical bandwidth [7], to implement loss-less high-speed systems [8]. They are available in both C-band and O-band. When designed in O-band, they are commonly used in coarse wavelength-division multiplexing (CWDM) systems [9]. Manifold researches have shown their role in optical communication networks. Schmuck et al. [10] presented SOA performance in open metro-access to create a transparent and reconfigurable optical ring network; in [11], they reported the importance of SOA efficiency in PON-applications as a booster and pre-amplifier. Ramírez et al. [12] described the essential role of SOA to extend the link power budget in short reach networks in an 8 × 50 Gb/s WDM system. SOAs are also considered as a substantial component for ultrafast all-optical signal processing devices, as wavelength converters [13], all-optical gates [14], and optical demultiplexers [15]. Recently, on-chip SOAs attracted a lot of attention due to their large-scale integration capability, low cost, and offer of more compact and smaller devices with low losses. Over the last few years, SOA integration technology has significantly been developed. Different technologies have been demonstrated for SOAs use in photonic chips, either by hybrid or by monolithic integration solution. Developing SOA is challenging on both pure InP platforms and hybrid III-V-on-silicon platforms. Specific coupling/bonding techniques are required such as flip-chip bonding [16, 17], die-to-wafer bonding [18], edge coupling [19], spot size converter (SSC) [20] and chip-to-chip butt coupling [21]. Furthermore, SOA monolithically integrated with photodiodes has been proposed as a promising technique for low-cost, high-speed, high-sensitivity SOA_PIN receivers [22, 23]. The recent demonstration of SOA heterogeneous integration on a silicon substrate by direct bonding of InP-based active region to the substrate introduced new possibilities for advanced PICs in the high wavelength areas [24]. More recently, a U-bend design has been proposed for III-V gain devices, such as SOAs, for simplifying the butt-coupling between the III-V chip and silicon-on-insulator photonic circuit [25]. All these techniques have been proposed to enable the SOA implementation on-chip for different applications. SOAs can operate either in the linear regime, as reach-extender amplifiers in gigabit passive optical networks (GPON) [26], or in the nonlinear regime, as an all-optical switching [27, 28]. SOA-based chip-level applications are not limited to optical communication systems but also include medical sensors, industrial and environmental monitoring [29], and nonlinear optics [30]. Depending on application requirements and chip fabrication constraints, the SOA working operation such as design may be very different. One of the most significant parameters is the gain value. SOAs can be designed to operate at high gain, as required on the receiver side to increase sensitivity [11], as well as at low gain/high output saturation power, as needed at the transmitter side in order to increase the lunch power [31]. Generally, chip-scale tunable lasers utilize a booster SOA. In integrated microwave photonics, a high gain SOA is required in combination with a tunable comb laser [32] or tunable ring-resonator-based lasers [33, 34]. In Ref. [35], a vertical-cavity surface-emitting laser (VCSEL) is co-packaged with a high gain SOA for OCT and LiDAR applications. Device parameters (e.g., active region dimensions and gain) and system parameters (e.g., input signal extinction ratio (*ER*)) have both a strong influence on SOA operation. Directly modulated lasers

(DMLs), typically used in low-cost transmission systems, have limited extinction ratio, which affects SOA dynamics [36] and decreases the signal quality. To solve this issue, the external light injection has been suggested by Lu et al. [37] for a 10 Gb/s optical transport system based on VCSELs and high gain SOAs. They used this technique to increase the SOA saturation power. By the increment of the injected light, they kept the SOA in the linear operation regime. In this paper, we demonstrate the possibility of using a low gain SOA after a low *ER* transmitter, such as DMLs, without any additional component. SOAs have also been indicated for compensating on-chip losses [38]. Utilizing SOA with high gain for compensating PIC's component losses has already been proposed for up to 164 optical components on a fully integrated free-space beam steering chip [39]. In the above-mentioned SOA-based PICs the quality factor of the amplified signal is not estimated. Assessing the quality of the optical signal is of paramount importance in the design of an optical system. Knowing the SOA impact on the signal quality during the design phase can remarkably reduce the number of fabrication runs, leading to significant cost savings. For access network applications, the SOA influence on the Q-factor has been theoretically and experimentally evaluated for finding the input power dynamic range (IPDR) as a function of bitrate, wavelength, and bias current [40]. That model is based on an empirical tolerance factor, a lower limit of the IPDR due to a low OSNR, and an upper limit due to gain saturation.

In this chapter, we propose a new analytical model for the SOA impact on the signal quality factor by taking into account the effects of gain compression, ASE noise, signal extinction ratio (*ER*), and receiver filter shape and bandwidth. The model assumes single-channel signal transmission, with conventional on-off key (OOK) non-return-to-zero (NRZ) modulation. The receiver is assumed to sample the signal at half of the bit time, as in all practical implementations. The proposed model provides two analytical equations for the Q-factor in the worst and best system operating conditions. For the analytical model proposed in this investigation, we considered some SOA specifications, such as active region dimensions and several effects that influence SOA performance, intending to estimate the Q-factor of an SOA performance before its actual integration into the system. The model can be used as a guideline to design a proper SOA in PIC or as a stand-alone device. The proposed model provides a tool for optimizing the SOA design. We apply the model considering two specific examples of SOA integrated into photonics interconnection systems or as a stand-alone device. The numerical model we used is essentially similar to that described in [41]. The steady-state and dynamic analyses are performed numerically for the SOA design using a parabolic gain model. The dynamic model used to simulate the SOA behavior is based on the numerical solution of the rate equation for carrier density, for a multi-section SOA. To solve the rate equations of each section, a standard ordinary differential equation (ODE) is used. The study is here limited to intensity variation neglecting the chirp effect that may be included in a more complete description when complex modulation formats are used in the system as well as long fiber span. We investigated the design of a low-gain SOA device that can be utilized as an on-chip loss compensator, or as a booster for a DML, through the analytical model and numerical simulations for various operating conditions. We estimate performance in terms of quality factors. Finally, the model is experimentally validated through a 10 Gb/s OOK WDM transmission system including a commercial SOA. We further discuss the extension of the model to higher-order modulation formats. In the following sections of the chapter, the proposed model is described in detail. In Section 2, the model is introduced together with its main parameters. Section 3 is divided into three subsections where gain compression, extinction ratio, and receiver filtering effects are modeled, respectively. Experimental setup and validation are described in Section 4. Section 5 shows two design examples and possible model extensions. Finally, conclusions are summarized in Section 6.

2. Model

The system under investigation is described through the block diagram in **Figure 1a**.

An optical transmitter generates the signal $P_{\text{in}}(t)$ at the wavelength λ as a sequence of binary symbols:

$$P_{\text{in}}(t) = \sum_{n=-\infty}^{+\infty} b_n p(t - nT_b) \tag{1}$$

where $b_n \in \left[P_{\text{in}}^L, P_{\text{in}}^H\right]$ are the transmitted bit power [mW], and $p(\mathrm{t})$ is an ideal rectangular pulse whose time duration is equal to T_b. When the probability of ones and zeros is equal, the average input power is $\bar{P_{\text{in}}} = \left(P_{\text{in}}^H + P_{\text{in}}^L\right)/2$, and the extinction ratio is defined as $ER \triangleq P_{\text{in}}^H / P_{\text{in}}^L$. The signal is amplified through an SOA that provides a gain, $G(t)$, depending on the SOA physical parameters, such as the driving current I_{dc}, and the input signal.

$$P_{\text{out}}^{\text{SOA}}(t) = P_{\text{in}}(t)G(t) \tag{2}$$

In the hypothesis of λ corresponding to the SOA gain peak, the saturated SOA gain must satisfy the equation [26]:

$$G(t) = G_0 \exp\left(-\frac{G(t)-1}{G(t)}\frac{P_{\text{out}}^{\text{SOA}}(t)}{P_{\text{sat}}}\right) \tag{3}$$

being G_0 the unsaturated gain (small-signal gain) and P_{sat} the saturation power. At the same time, SOA generates noise $n(\mathrm{t})$, modeled as a white Gaussian noise, whose power spectral density is a function of the gain and the population-inversion factor $n_{\text{sp}}(\mathrm{t})$, such as Eq. (4).

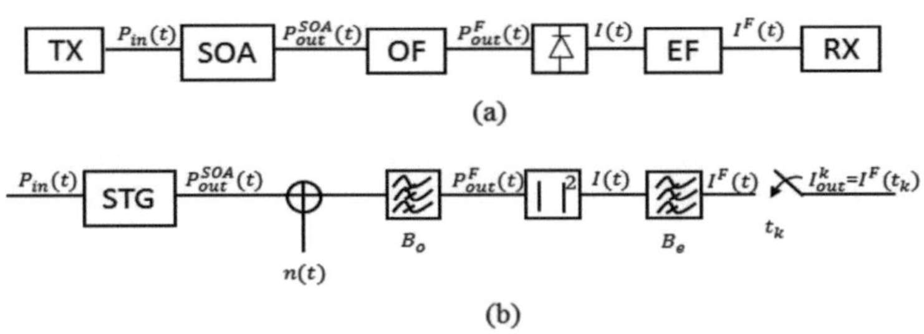

Figure 1.
(a) Block diagram, and (b) System representation. OF: optical filter, EF: electrical filter, and STG: saturated gain.

$$N_{\mathrm{ASE}}(t) = 2n_{sp}(t)\frac{hc}{\lambda}[G(t) - 1] \tag{4}$$

where h is the Plank constant and c *is* the speed of light in the vacuum. After amplification, in order to reduce the impact of the noise, an optical band pass filter is included with response time $h_o(t)$ and 3-dB bandwidth B_o so that

$$P^{F}_{\mathrm{out}}(t) = P^{\mathrm{SOA}}_{\mathrm{out}} * h_o(t) \text{ and } P_{\mathrm{ASE}}(t) = N_{\mathrm{ASE}}(t) * h_o(t) \tag{5}$$

By neglecting thermal and quantum noise, the photodetected current is:

$$I(t) = R\left|\sqrt{P^{F}_{\mathrm{out}}(t)} + \sqrt{P_{\mathrm{ASE}}(t)}\right|^2 = R\left[P^{F}_{\mathrm{out}}(t) + 2\sqrt{P^{F}_{\mathrm{out}}(t)P_{\mathrm{ASE}}(t)} + P_{\mathrm{ASE}}(t)\right] \tag{6}$$

Being R [A/W] the photodetector responsivity. The first term in Eq. (6), represents the signal, whereas both the other two terms are noise contributions. At the receiver input, an electrical filter with bandwidth B_e is considered, giving a bandlimited new signal including the additive noise

$$I^{F}(t) = I(t) * h_e(t) \tag{7}$$

where $h_e(t)$ represents the electrical filter impulse response. It is usual to evaluate the performance of an optical communication system at the optimal sampling time. Here, for the sake of simplicity and for considering a realistic clock-data recovery circuit, a fixed sampling time at $T_b/2$ will be considered, so that the output samples take the values

$$I^{k}_{\mathrm{out}} = I^{F}(kT_b/2) \tag{8}$$

Each sample represents a distorted binary symbol with an additional white Gaussian noise whose variance depends on the instantaneous noise contribution as described in Eq. (6). When an OOK bit stream as represented in **Figure 2a** is propagated through the system, the photo detected current will assume the generic behavior of **Figure 2b** including those distortions that can be ascribed to the saturated gain and the additional noise. By neglecting the effect optical and electrical filters, it is possible to identify the highest and lowest values for samples corresponding to each input bit-level providing:

$$I^{L}_{b} = \min\left\{I^{k}_{\mathrm{out}}\right\}, I^{L}_{w} = \max\left\{I^{k}_{\mathrm{out}}\right\} \tag{9}$$

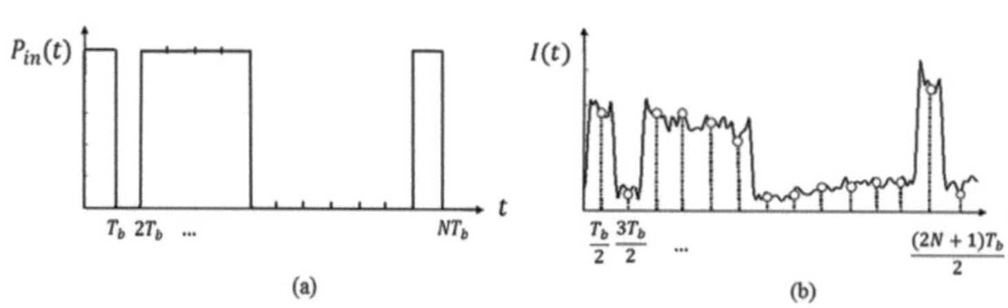

Figure 2.
(a) Input signal example, (b) distorted output. Sampled values are marked at $T_b/2$ interval and they are shown with circles.

when P_{in}^{L} is transmitted, and

$$I_b^H = \max\left\{I_{\text{out}}^k\right\}, I_w^H = \min\left\{I_{\text{out}}^k\right\} \tag{10}$$

when P_{in}^{H} is transmitted.

In the simple case of linear gain, $I_b^L = I_w^L = I_L$ and $I_b^H = I_w^H = I_H$. For what concerns noise contributions to the same samples (at $T_b/2$), we assume a white Gaussian noise whose variances, associated with the input bit P_{in}^L and P_{in}^H , can be written as:

$$\sigma_{L,H}^{k} = R\left\{\left[2P_{\text{out}}^{F}(kT_b/2)P_{\text{ASE}}(kT_b/2)+P_{\text{ASE}}^{2}(kT_b/2)\right]B_e/B_o\right\}^{1/2} \tag{11}$$

Noise variances assume four discrete values in correspondence of best and worst cases for low and high input power that will be indicated with $\sigma_{L,H}^{b,w}$.

As it is shown in **Figure 3**, when the optimal threshold is used for bit decision [42] and bit one and zeros have the same probability, the bit error rate (BER) can be found, in the two cases, by the highlighted areas so that:

$$\text{BER}_w = \frac{1}{2}\frac{1}{\sigma_L^w\sqrt{2\pi}}\int_{I_w^{th}}^{\infty}\exp\left[-\frac{1}{2}\left(\frac{I-I_w^L}{\sigma_L^w}\right)^2\right]dI + \frac{1}{2}\frac{1}{\sigma_H^w\sqrt{2\pi}}\int_{-\infty}^{I_w^{th}}\exp\left[-\frac{1}{2}\left(\frac{I-I_w^H}{\sigma_H^w}\right)^2\right]dI \tag{12}$$

$$BER_b = \frac{1}{2}\frac{1}{\sigma_L^b\sqrt{2\pi}}\int_{I_b^{th}}^{\infty}\exp\left[-\frac{1}{2}\left(\frac{I-I_b^L}{\sigma_L^b}\right)^2\right]dI + \frac{1}{2}\frac{1}{\sigma_H^b\sqrt{2\pi}}\int_{-\infty}^{I_b^{th}}\exp\left[-\frac{1}{2}\left(\frac{I-I_b^H}{\sigma_H^b}\right)^2\right]dI \tag{13}$$

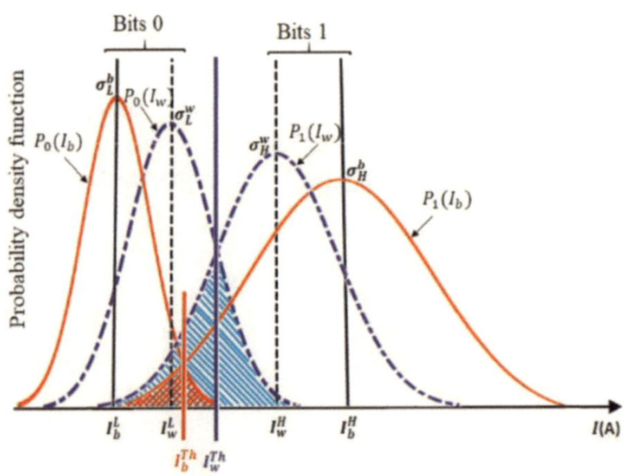

Figure 3.
Probability density function $P_{0,1}(I_b)$ corresponding to the best case at mean one I_b^H and zero I_b^L levels (with a standard deviation σ_H^b and σ_L^b) with the optimal threshold for bit decision I_b^{Th} in red and probability density function of the worst $P_{0,1}(I_w)$ at mean one I_w^H, and zero I_w^L levels (with σ_H^w and σ_L^w) with the optimal threshold for bit decision I_w^{Th} in light blue. The hatched red area determines the BER_b for the best case while the hatched blue area represents the BER_w for the worst case.

When threshold is optimized,

$$\mathrm{BER}_w = \frac{1}{2}\mathrm{ercf}\left(\frac{Q_w}{\sqrt{2}}\right) \text{ and } \mathrm{BER}_b = \frac{1}{2}\mathrm{ercf}\left(\frac{Q_b}{\sqrt{2}}\right) \tag{14}$$

where,

$$Q_w = \frac{I_w^H - I_w^L}{\sigma_H^w + \sigma_L^w} \text{ and } Q_b = \frac{I_b^H - I_b^L}{\sigma_H^b + \sigma_L^b} \tag{15}$$

In the linear case,

$$Q_w = Q_b = Q = \frac{I_H - I_L}{\sigma_H + \sigma_L} \tag{16}$$

In the simulation and experimental results, the quality factor (eye-opening) is considered based on the signal received by the decision circuit that samples at the decision instant ($T_b/2$) as described in Eq. (8). The sampled values fluctuate from bit to bit around an average value I^H or I^L, depending on the bit logical value in the bit stream [42]. The decision circuit compares the sampled value with a threshold value Th. In the analytical model, we estimate a boundary condition for the SOA quality factor based on the highest and lowest sampled values corresponding to each input bit-level. The sampled values located around an average value $I_b{}^H/I_b{}^L$ or $I_w{}^H/I_w{}^L$ provide boundary conditions Q_b and Q_w, respectively, as described in Eqs. (9) and (10). The best and worst quality factors correspond to linear and nonlinear SOA operation as it will be clarified hereafter.

Figure 4 shows, as an example, 10 Gb/s NRZ signal eye diagrams when passing an SOA operating in nonlinear (top) and linear(bottom) regimes. On the left-hand side, analytical probability density functions (pdf) are reported in the two boundary cases; to the right, simulated pdfs are shown for comparison. SOA parameters are given in Section 4. When the input signal power is high enough (–12dBm in the example), SOA suffers for nonlinearities and the analytical model estimates a Q-factor to be $Q_w =$ 13.3 dB. By lowering the power down to –19 dBm, SOA is almost linear and the model returns $Q_b = 12.2$ dB. For both cases, the extinction ratio of the rectangular input signal is 6 dB. In the same operating conditions, simulated values of Q-factor, as shown in **Figure 4** right, are 13.8 and 11.9 dB for the nonlinear and linear cases, respectively, which are in good agreement with the analytical ones. The pdfs on the left, with the sampling time at the beginning of the bit, show the separated high and low levels for best and worst cases to assess the analytical Q_b and Q_w with the optimal thresholds ($I_b{}^{Th}/I_w{}^{Th}$). Experimental validation is provided in Section 4. When the optical input power exceeds the linear regime, the two bound values strongly depend on several parameters, including SOA gain saturation and recovery time, as well as input signal ER, bit pattern and rate, filters bandwidth, and shape.

Although the model assumes single-channel transmission, it is even valid for multichannel systems. In fact, if the linear operation description is trivial, in the nonlinear case, the worst estimation is given when all the WDM channels are assumed to be synchronous with the same data stream. Since this analytical model is based on the gain variation in the SOA, any effect that has an impact on that, including nonlinear gain modulation, contributes to Q-factor degradation. For instance, consider the SOA performance in the WDM system (as in line amplifier) for 12 $\times$25 Gb/s

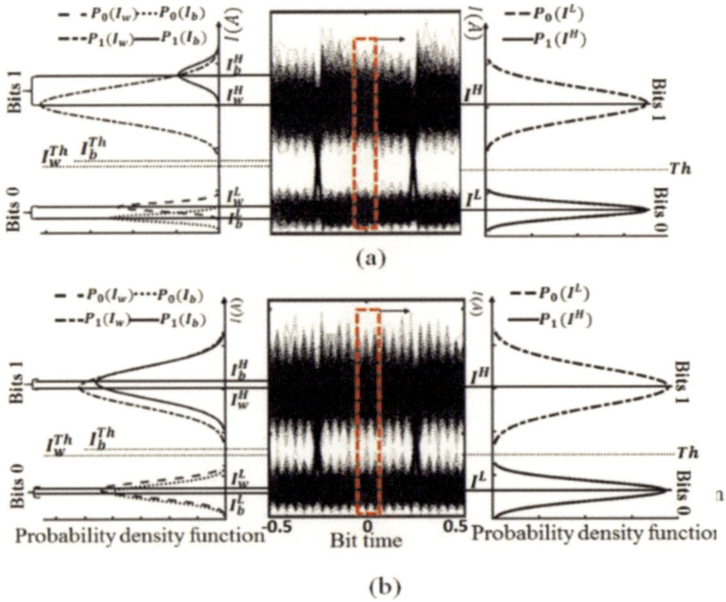

Figure 4.
Normalized eye diagrams and probability density functions (pdf) for SOA operation in nonlinear (a), and linear (b) regimes, left side: shows pdfs considering Q_b and Q_w, and right side: shows pdfs considering actual Q-factor.

NRZ transmission with 200 GHz channel spacing with the power per channel equal to –12 dBm and extinction ratio of 15 dB. In this case, the total power is –1.2 dBm (corresponding to –12 dBm per channel). Using the model for a single channel with –1.2 dBm, provides the estimated boundary conditions of the quality factor for all 12 channels. In the following different impairments will be studied and analytical expressions provided.

3. Performance estimation

3.1 Saturated gain

In order to isolate the effect of saturated gain, the following assumptions have been made:

1. $P_{\text{in}}^{L} = 0 \Rightarrow \text{ER} = \infty \Rightarrow I_b^L = I_w^L = I_L = 0,$

2. Ideal optical and electrical filters so that B_e/B_o=1.

When a binary pattern with an infinite extinction ratio is propagated through an SOA, each bit one is instantaneously amplified with a gain that depends on the previous bit sequences and can vary between the small-signal linear gain G_0 and the minimum value represented by the saturated gain G_s obtained for a constant input power P_{in}^{H}. The former case occurs when the pattern preceding bit one is composed of a number of consecutive zeros large enough for a complete gain recovery. This number depends on the carrier lifetime τ_c and the bit time T_b as it is shown in [43]. The highest power at the

τ_c/T_b	1.3	5.2	13.1	26.3
$R_b(Gb/s)$	2.5	10	25	50
P_{in}^{sat}=-13dBm P_{in}=-12dBm	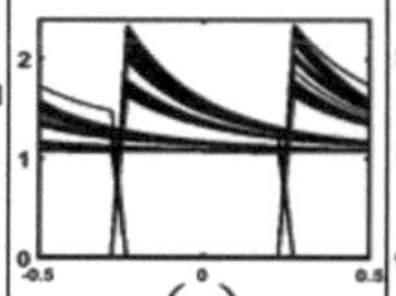 (a)	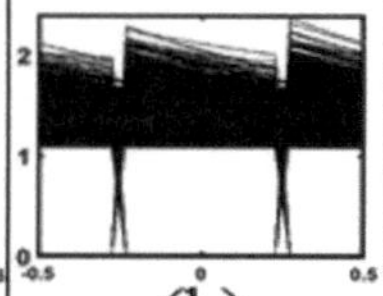 (b)	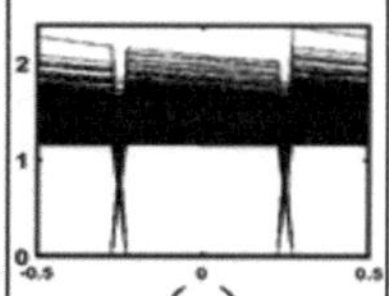 (c)	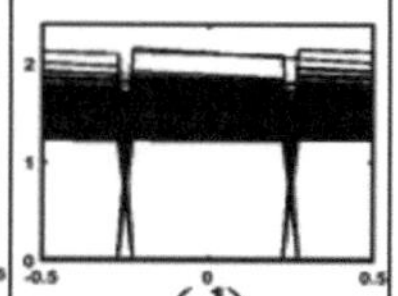 (d)

Figure 5.
Simulated eye diagrams (Power (mW)) for various bit rates at 1555 nm at I_{dc} *= 70 mA, and at the constant input power (–12 dBm). Eye diagrams show the gradual changes in the Q-factor boundaries due to the ratio* (τ_c/T_b).

sampling instant is obtained when $\tau_c \gg T_b$ and can be approximated with $P_{\text{in}}^H G_0$. Where for the $(\frac{\tau_c}{T_b} \geq 8)$, the input power is considered as its average value [43]. On the other hand, the lowest output power is obtained for steady-state operation when SOA reaches saturation and can be expressed by $P_{\text{in}}^H G_s$. SOA gain recovery depends on the relation between τ_c and T_b. **Figure 5** clarifies the SOA gain dynamic operation in the case of constant input power (–12 dBm) for different $\frac{\tau_c}{T_b}$ ratio values.

When the bit rate is low, T_b is comparable with τ_c and SOA gain can recover to its highest value (small-signal gain) within a few bit times when bit is zero. Each bit one, following one or more-bit zeros, experiences a gain that is high at the beginning of the bit time but it quickly decreases to the saturated value. Saturation depends on input power as well as bit distribution and modulation extinction ratio. As the bit rate increases, both gain saturation and recovery gradually flatten and SOA becomes transparent to the bit pattern if that does not include very long sequences of bit zero. The eye diagrams in **Figure 5** do not include ASE noise.

Boundary conditions can be easily determined from Eq. (15), such as:

$$Q_w = \frac{I_w^H}{\sigma_H^w + \sigma_L^w} = \frac{P_{\text{in}}^H G_s}{P_{\text{ASE}}^0 + \sqrt{\left(P_{\text{ASE}}^s\right)^2 + 2P_{\text{ASE}}^s P_{\text{in}}^H G_s}} \tag{17}$$

and,

$$Q_b = \frac{I_b^H}{\sigma_H^b + \sigma_L^b} = \frac{P_{\text{in}}^H G_0}{P_{\text{ASE}}^s + \sqrt{\left(P_{\text{ASE}}^0\right)^2 + 2P_{\text{ASE}}^0 P_{\text{in}}^H G_0}}, \tag{18}$$

where P_{ASE}^s and P_{ASE}^0 are the ASE powers calculated through Eq. (4), in the saturated and linear conditions respectively. **Figure 6a** shows an ideal rectangular NRZ_OOK bit pattern with infinite *ER* launched through the SOA. The behavior of the detected signal pattern by neglecting SOA_ASE noise is shown in **Figure 6b**. While **Figure 6c** shows the effect of the SOA_ASE noise on the signal pattern as demonstrated in Eq. (6). The sampled values are shown with the dotted lines and the I_b^H and I_w^H marked by dash-dotted and dashed lines, respectively. In this pattern stream, the highest power at the sampling instant of bits one is obtained from the sampling of the bit one that comes after a long sequence of bits zero, marked with an empty circle in

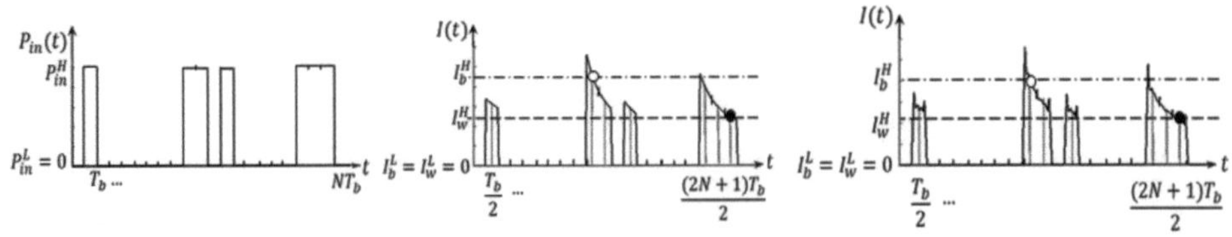

Figure 6.
(a) Input signal example with infinite extinction ratio. Distorted output: (b) without ASE, (c) with ASE effect. Sampled values are shown with the dotted lines. The highest and the lowest output power at the sampling are marked by the dash-dotted and dashed lines, respectively. At the sampling instant of bits one, the highest power is marked by an empty circle, the lowest power is marked by a filled circle.

Figure 6b and **c**. On the other hand, the lowest power obtained from the sampling of the bit one comes after several consecutive ones, marked with a filled circle. This fact shows that the signal distortion due to the SOA nonlinearity is pattern-dependent, where each bit one is amplified with a gain that depends on the preceding bit sequences. As is shown in **Figure 6b**, when bit one arrives, providing its pattern preceding composed of a long sequence of zeros, the output power achieves its highest level. The output power reduces for the next arriving bits one, which is a decaying function of time starting from an initial value of power [43], which can be attributed to the carrier density depletion. The gain impact on signal quality in both linear and nonlinear regimes of SOA operation was investigated using Eqs. (17) and (18). By maintaining the device injection current constant, when the input signal power is low, SOA provides linear gain and high ASE. Whereas, gain and ASE reduce when the input power increases, thus affecting the signal Q-factor. **Figure** 7a shows the simulated gain as a function of input power for a wavelength of 1555 nm at variable injected currents such as 200, 92, and 70 mA corresponding at 30.9, 17.2, and 15.3 dB small-signal gain, respectively. Simulation results have been obtained using Eq. (3) under static operation conditions [41] for an SOA with 1800 μm-length and 0.38 μm-width active region. The arrows show saturation input power $P_{\text{in}}^{\text{sat}}$, where the gain is reduced by 3-dB, for various bias currents. The analytical results of the Q_b and Q_w for the 200 mA driving current are presented as a function of gain in **Figure 7b**, reminding that the decreasing gain is due to the increasing input power. It is clear from the results that the signal quality increases while the gain decreases. The reason for this behavior is the ASE impact reduction on the signal stream at lower gain. Since ASE is a

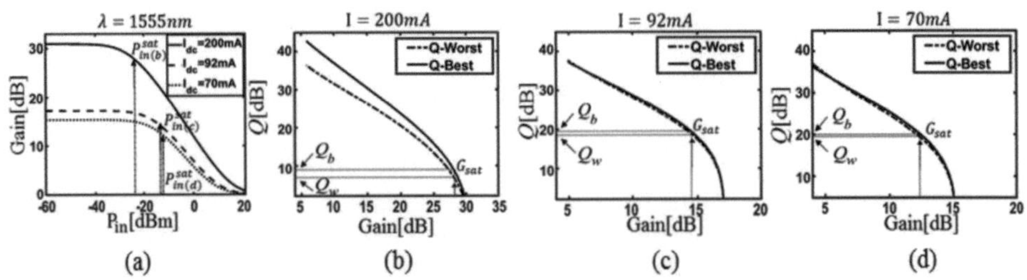

Figure 7.
(a) Simulated gain as a function of input power at I_{dc} applied in (b), (c), and (d). Arrows show the corresponding $P_{\text{in}}^{\text{sat}}$ at each I_{dc}. Quality factor as a function of SOA gain at different drive current: (b) I_{dc}=200 mA, G_o=30.9 dB and Q at $P_{\text{in}}^{\text{sat}}$: Q_b=8.8 dB, Q_w=6.9 dB (c) I_{dc}=92 mA, G_o=17. 2 dB and Q at $P_{\text{in}}^{\text{sat}}$: Q_b =19 dB, Q_w =18.3 dB (d) I_{dc}=70 mA, G_o=15.3 dB and Q at $P_{\text{in}}^{\text{sat}}$: Q_b=20 dB, Q_w =19.4 dB. The Q_b and Q_w at the G_{sat} marked by the dotted lines. G_{sat} is the 3dB gain saturation at the saturation input power $P_{\text{in}}^{\text{sat}}$.

function of the gain, the signal performance is limited by ASE noise at high gain (low input power), while at a low gain (high input power), the signal performance is limited by patterning. In this case, the signal distortion at high input power has not severe effect on signal performance due to infinite ER assumptions. Moreover, the impact of bias current on the device gain and consequently on signal quality was investigated. The injected current was reduced from 200 mA down to 92 mA (17.2 dB gain) and 70 mA (15.4 dB gain).

Results are shown in **Figure 7c** and **d**, respectively. In the linear regime, as gain remains almost constant ($\sim G_0$) the ASE contributions on I_w^H and I_b^H will be equal, which leads to identical results for both best and worst quality factor approximation ($Q_b = Q_w$). Whereas, as the gain starts to saturate (nonlinear regime), the impact on signal quality gets intense at high injected current (200 mA), which translates into a small decrease of the Q_w value (**Figure 7b**). The signal quality at a saturation input power $P_{\text{in}}^{\text{sat}}$ improves at the lower injected current, where the gain value is smaller and carrier lifetime is longer. Therefore, whenever gain decreases, either by reducing injected current or increasing input power, the signal quality increases. It is worth mentioning that reducing injected current leads to decreasing G_0 and relatively decreases G_s, while increasing input power leads to decreasing G_s, but G_0 remains unchanged (without consideration ER effects). Accordingly, both small-signal gain and saturated gain have an impact on Q-factor. Indeed, the contribution of G_0 and G_s in Eqs. (17) and (18) are also through the $P^0{}_{\text{ASE}}$ and $P^S{}_{\text{ASE}}$, respectively. In another way, the ASE noise affects the optical signal-to-noise ratio (SNR), particularly at low input power where ASE is higher. Thus, at high gain, high ASE, the signal performance degrades (lower Q-factor). We noticed that, at low bias current and high input power, the obtained Q_w value is slightly higher than the Q_b. We attributed this increase of Q_w to the reduction of the ASE impact on I_w^H, since I_w^H corresponds to a lower gain, while I_b^H is given by the maximum value of the SOA gain, i.e., G_0.

It should be noticed, in this case, that for isolating the gain effect, we ignored all other effects on the signal quality that led to obtaining quality factors higher than realistic cases.

3.2 Extinction ratio

One of the most significant parameters that should be considered for a launched signal through an SOA is the signal ER, as it affects the amplified signal quality. When a binary pattern with finite ER propagates through an SOA, amplification occurs on both input bit levels. So that, not only bits one but also bits zero instantaneously experience amplification with the gain that depends on their former bit sequences (**Figure 8**). The highest power for bit zero is obtained when the pattern preceding bit zero is composed of several consecutive zeros, large enough for complete gain recovery, and can be approximated by $P_{\text{in}}^H G_0/\text{ER}$. Furthermore, the lowest power is obtained when SOA reaches saturation, and the pattern-preceding bit zero is composed of several consecutive ones. The lowest power for bits zero can be approximated by $P_{\text{in}}^H G_s/\text{ER}$. Following Eqs. (17) and (18), the best and worst quality factors in the case of limited extinction ratio will define as:

$$Q_w = \frac{I_w^H - I_w^L}{\sigma_L^w + \sigma_H^w} = \frac{P_{\text{in}}^H G_s - \left[\left(P_{\text{in}}^H/\text{ER}\right)G_0\right]}{\sqrt{\left(P_{\text{ASE}}^0\right)^2 + 2P_{\text{ASE}}^0\left(P_{\text{in}}^H/\text{ER}\right)G_0} + \sqrt{\left(P_{\text{ASE}}^s\right)^2 + 2P_{\text{ASE}}^s P_{\text{in}}^H G_s}} \tag{19}$$

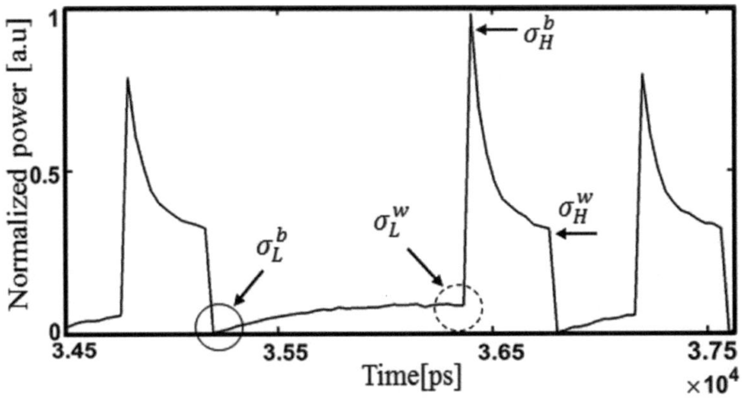

Figure 8.
Part of the simulated distorted output of an ideal rectangular input with an extinction ratio of 10 dB. Arrows show best and worst cases for low and high noise variances. Solid and dashed circles indicate the amplified zero level results of saturated gain and full recovery gain respectively.

and,

$$Q_b = \frac{I_b^H - I_b^L}{\sigma_L^b + \sigma_H^b} = \frac{P_{\text{in}}^H G_0 - \left[\left(P_{\text{in}}^H/\text{ER}\right)G_s\right]}{P_{\text{ASE}}^s + \sqrt{\left(P_{\text{ASE}}^0\right)^2 + 2P_{\text{ASE}}^0 P_{\text{in}}^H G_0}} \tag{20}$$

In **Figure 8**, part of the simulated output signal pattern of an ideal rectangular input signal, with an extinction ratio of 10 dB, is represented. Four discrete values of best and worst case, for low and high noise variances, are marked by the arrows. Referring to **Figure 8**, it is evident, when the ER is finite, the low ('zero') level will be amplified too. Besides, its output power is variable due to gain variation from saturated gain (G_s) to full recovery gain (G_0). Given that, the eye-opening of the signal is affected by patterning and, as a consequence, the quality factor reduces. When SOA operates in the nonlinear regime, transmitting a long sequence of bits leads to gain compression due to carrier depletion, and gain achieves its lowest value. While, after transmitting a long sequence of bits zero, the SOA's carrier density, and therefore its gain, recover to their maximum achievable values. Following this, when bit one arrives, the output achieves its maximum I_b^H, the carrier depletion starts, subsequently, the output starts to decay (**Figure 8**).

Under static operation conditions, maximum SOA gain can be found as a function of the average input power $\bar{P}_{\text{in}} = \left[P_{\text{in}}^H + \left(P_{\text{in}}^H/\text{ER}\right)\right]/2$ for low ER. For high ER values (ER_h), the maximum gain value assumes the small-signal value of G_0, but when the ER is decreased (i.e., ER < 10 dB as in practical systems) to lower values (ER_l), the maximum gain is indicated as $G_l \leq G_0$. When ER is smaller than 10 dB, the fully recovered gain depends on the input power and ER (ER < 10 dB). Exploiting Eq. (3), the SOA gain (G_{l0}) can be found as a function of the average input power for ER < 10 dB (Eq. (22)). Gain equations, hence, will take the form:

$$G_0 = \exp\left(gL\right) \tag{21}$$

$$G_{l0} = G_0 \exp\left\{\left(1 - G_{l0}\right)\frac{\left[P_{\text{in}}^H + \left(P_{\text{in}}^H/\text{ER}\right)\right]}{2P_{\text{sat}}}\right\} \tag{22}$$

$$G_l = \frac{G_0 + G_{l0}}{2} \tag{23}$$

$$G = G_0 \exp\left[(1 - G)\frac{P_{\text{in}}}{P_{\text{sat}}}\right] \tag{24}$$

where g is the net gain per unit length and L is the active region length. Being $P_{\text{sat}} = h\nu\sigma_m/\Gamma\tau_c a$ the saturation power of the SOA, where ν is the optical frequency, Γ is the optical confinement factor, a is the differential gain coefficient, and $\sigma_m = W \cdot d$ is the active region cross-section, where W and d are its width and thickness, respectively. Eq. (22) can be solved numerically to obtain the full recovery gain G_l at ER_l (Eq. (23)). Thus, the average of G_{l0} and SOA small-signal gain (G_0) results in an estimated full recovery gain value for low ERs. This value is used as full recovery gain in Eqs. (19) and (20) when ER is low. Results are shown in **Figure 9**, where the maximum gain is calculated as a function of the input power for two ER values (ER_h = 20 dB (G_0 curve (a)), and ER_l = 6 dB (G_l curve (b))) and compared to the gain saturation curve (c). As it is illustrated in curve (b) of **Figure 9**, the fully recovered gain (G_l) for ER_l = 6 dB changes with input power. So that, at low input power, G_l (curve (b)) is close to the small-signal gain G_0 (curve (a)), but by increasing input power, it gets spacing, where the distance depends on the input power and ER_l. When the ER is low (ER<10 dB), the impact of the amplified bits zero at any input power level, high or low, on signal quality is significant. For instance, at an input power of –12 dBm, the saturated gain is equal to 12.3 dB for both cases, high and low ERs, as is shown in **Figure 9** (marked by a star). However, the value of the full recovery gain is related to the value of the input signal ER. Exploiting Eqs. (22) and (23) for ER = 6 dB, full recovery gain at the saturation input power (–12 dBm) obtained equal to 14.1 dB (marked by the square in **Figure 9**), which is 1.2 dB less than the full recovery gain at

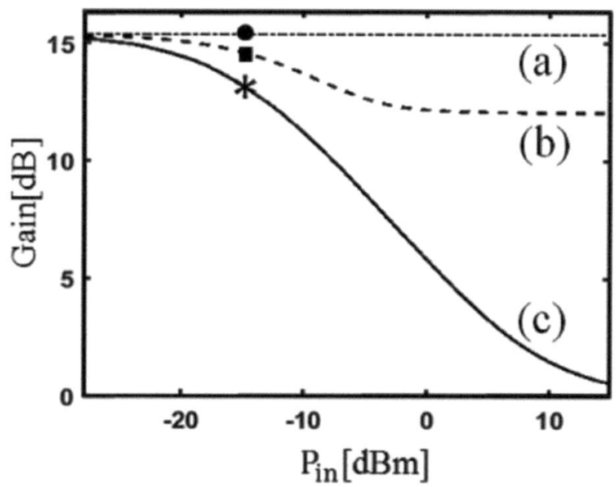

Figure 9.
Saturated amplifier gain G as a function of the input power for 15.3 dB of the small-signal gain. (a) Full recovery gain for input signals with high ER >10 dB. Obtained by Eq. (21). The value of full recovery gain is constant and it is equal to small-signal gain for any input power and ER>10 dB, and (b) Full recovery gain for input signals with low ER $\leq$ 10. Obtained by Eqs. (22) and (23). The value of full recovery gain changes by changing ER (for ER $\leq$ 10) and it depends on the value of input power. The graph in (b) Obtained for ER = 6 dB. (c) Saturated amplifier gain, which obtains equally for both high and low ER using Eq. (24). (I_{dc} = 70 mA, λ = 1555 nm). The fully recovered gain for ER = 6 dB and ER = 20 dB and the saturated gain, for both cases, at –12 dBm input power are marked by square, filled circle, and star, respectively.

high ER (marked by a filled circle in **Figure 9**) achieved by Eq. (21). Employing these gains in Eqs. (19) and (20) lets us estimate the Q-factor. At confined ER, both bit levels (0s and 1s) are taking gain, and a number of the carriers will be used by P_{in}^{L}. The lower the *ER* (higher P_{in}^{L}), the more carriers will be taken by zero level power. This leads to reducing the number of carriers used by P_{in}^{H} and caused to change the value of full recovery gain, while at very high *ER* the carriers will be fully available for P_{in}^{H}.

Eventually, for low ER input signals, in Eqs. (19) and (20), G_l must be used instead of G_0. The analytical results of Q-factor behavior versus input power are represented in **Figure 10**. These results are obtained with various ERs for an SOA with a linear small-signal gain of G_0 = 15.3 dB and saturation input power of –13 dBm. At high ER (20 dB) and low launch powers, the Q_w and Q_b have similar behavior (as explained above), while at high input powers, they act differently. By increasing input power, the gain decreases, and consequently, ASE noise decreases, the Q_w and Q_b increase accordingly (**Figure 10a**). By decreasing, *ER* induces intensive patterning, thus the behavior of Q_w changes (**Figure 10b**) and it turns to a parabolic curve. In fact, in the linear regime, the ASE noise dominates, while in the nonlinear regime, patterning becomes the main signal degradation effect. Moreover, a further ER decrease introduces higher noise contribution in Q_w, which ends in reducing Q_w in linear operation area **Figure 10c** and **d**.

At low ER and high input power, where the Q_w flattened, the SOA reaches the transparency point, where the absorptions (losses) and emissions (gain) are identical within the SOA. Input power beyond this point drives the SOA below the transparency, where it will not be able to recover its gain. In practice, at this point (i.e., material gain transparency [44]), the value of the quality factor will be unmeasurable and useless. **Figure 11a** and **b** show the contour plot for the calculated Q_b and Q_w versus the input power and a large range of *ER*. The plot was obtained using our analytical model for an SOA with the small-signal gain of G_0 = 15.3 dB at the wavelength of 1555 nm and injected current of 70 mA. The given boundary condition is equal to the Q-factor of 15.6 dB, corresponding to a BER of 10^{-9}, and it is marked with the solid line. This enables the evaluation of the impact of the signal extinction ratio within different input powers on the amplifier performance. Eventually, the Q-factor is higher at the high extinction ratio signals, particularly on the high input power side. The SOA parameters used in our model have been derived from the characterization measurement of commercial SOA used in our experimental setup.

3.3 Filtering

As it is shown in **Figure 1b**, the transmission system includes two bandpass filters, the first one, in the optical domain, usually placed just before photodetector, whose role

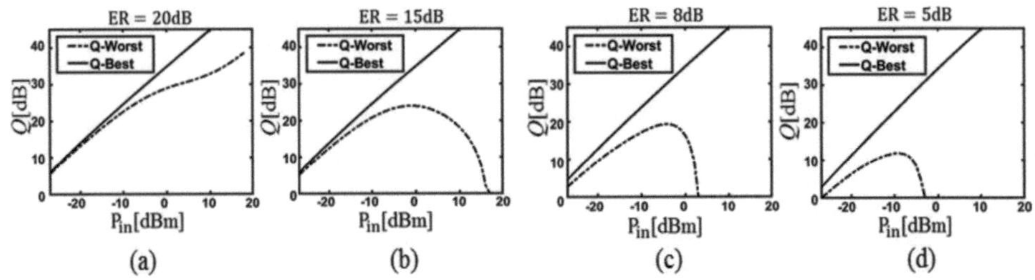

Figure 10.
Quality factor as a function of input power with, (a) ER = 20 dB, (b) ER = 15 dB, (c) ER = 8 dB, and (d) ER = 5 dB with small-signal gain G_0 = 15.3 dB.

DOI: http://dx.doi.org/10.5772/intechopen.103048

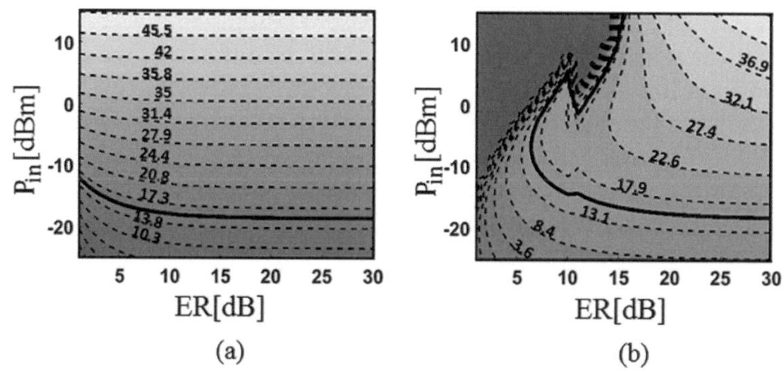

Figure 11.
Range of signal performance dependence on the input power and various ER for an SOA with an unsaturated gain of G_0 = 15.3 dB. Lines separate the boundary with different quality factors in the (a) best (Q_b) and, (b) worst (Q_w) cases. The solid lines mark the boundary with error-free amplification of a data stream where the Q = 15.6 dB corresponds to a BER of 10^{-9}.

is to remove out-band noise (such as ASE noise or cross talk due to adjacent channels). The second one in the electrical domain is generally matched to the transmission signal bandwidth to maximize the signal-to-noise ratio before the sampler. In a typical case of OOK NRZ transmission, a 4th order Bessel-Thomson filter with a bandwidth of 75% the bit rate is used. However, when nonlinearities affect the signal, such as SOA gain saturation, the matched filter does not provide optimal performance.

In this work, we performed analytical modeling of electrical filter effects on the Q-factor of the data streams and evaluated them through numerical simulation. At the first stage, we removed the SOA and optical filter. The effect of the electrical filter was considered on the received rectangular bit pattern. Then it was extended to investigate the electrical filter effects on the amplified bit pattern while SOA operated at the nonlinear regime. The numerical simulation of the filter effect on the data pattern and eye diagram of the rectangular signal is depicted in **Figure 12**. Applying the filter

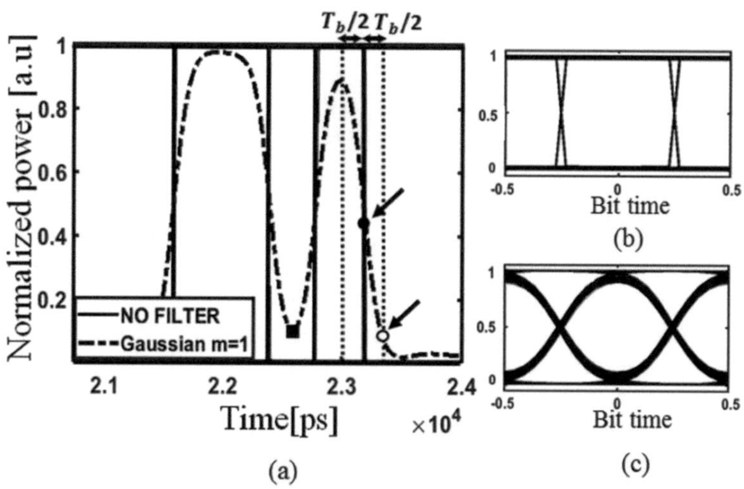

Figure 12.
(a) Simulated filter effect on rectangular data pattern, (b) eye diagram of unfiltered signal, and (c) filtered by G_1. Arrows show normalized filter value at the end of the duration of the bit one (0.5) and in the sampling instant of the bit zero (0.062). ISI effect on bit zero is marked with a square.

shapes the signal and affects the power of bits zero. Taking into account these facts, the effect of a Gaussian filter on signal quality was investigated. The performance of the Gaussian filter has also been compared with other conventional filters such as Super Gaussian and Raised Cosine (RC) filter. At the receiver input, an electrical Gaussian filter is included, where its 3-dB bandwidth is set to a frequency of 75% of the transmission data rate. The impulse response for the low-pass Gaussian filter, using Gaussian function, is defined as [45]:

$$h(t) = \exp\left(-t^{2m}/2\sigma^{2m}\right) \tag{25}$$

where σ is the filter bandwidth, and it can define based on the bit time ($\sigma = \alpha T_b$), and m is the filter order. Where $m = 1$ refers to the Gaussian filter, $m > 1$ refers to the Super Gaussian filter. Applying a first-order Gaussian filter (G_1) on the rectangular data stream, which would be applied to each symbol, makes each rectangular symbol becomes Gaussian-like. For the sake of simplicity, we normalized the signal power to unity. In this case, the normalized filter value (F) on the adjacent bit, when the bit is zero, can be calculated by taking the assumption of the filter value at the end of the duration of the preceding bit one be equal to 1/2, marked by a filled circle in **Figure 12a**. Considering the time interval from the second half of bit one to the first half of bit zero, the impulse response at $T_b/2$ would be $h(t - T_b/2) = 1/2$, using Eq. (25) the α, and subsequently, the filter bandwidth ($\sigma = 0.42\ T_b$) is obtained. The value of this filter in the sampling instant of the adjacent bit zero is $F_{G1} = 0.062$, marked by an empty circle in **Figure 12a**. We must notice that this filter value is valid when there is no Inter Symbol Interference (ISI) between two consecutive bits. The electrical filter can induce the ISI, where the bit gets smoothed out, and its energy spills over into the adjacent bits. The ISI causes a severe filter effect on bits zero and modifies the filter value on that, as it is shown in **Figure 12a** marked with a square. However, we demonstrate that using the filter value in the non-ISI case is enough to have a good estimation for the Q-factor. **Figure 12a** shows that one-level bits have different values. The reason for that can be found in the filter bandwidth. Implementing the same procedure, filter values at the adjacent bit sampling time for the second-order Gaussian (G_2) and Raised cosine (RC) filters are $F_{G2} = 0.183$ and $F_{RC} = 0.001$, respectively. The impulse response of the low-pass RC filter [46] is given by:

$$h(t) = \frac{\sin(t/T_b)}{t/T_b}\,\frac{\cos(\pi\alpha t/T_b)}{1-(2\alpha t/T_b)^2} \tag{26}$$

where $\alpha = 0.9$ is the roll-off factor. Due to the analytically obtained low filter value at the sampling instant of bits zero for RC ($F_{RC} = 0.001$), the RC is estimated to provide an open eye diagram. The effect of the G_2 and RC filters on eye-opening are simulated and depicted in **Figure 13**. The filter values obtained through the analytical model are validated numerically. As a further step, the filter effect on amplified signal is investigated analytically, and the results are evaluated through numerical analysis. In order to isolate the filtering effect, we assume that the filter acts on the received bit stream represented by a rectangular bit pattern with a very high extinction ratio (37 dB) propagated through an SOA. In this case, best and worst quality factors can be expressed as:

$$Q_w = \frac{I_w^H - I_w^L}{\sigma_L^w + \sigma_H^w}$$

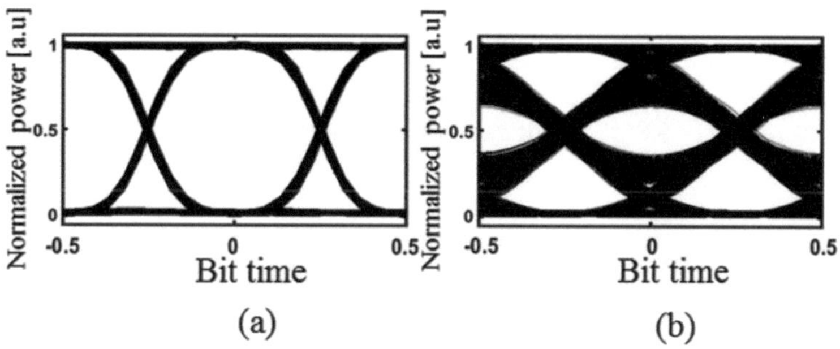

Figure 13.
Simulated eye diagrams of filtered rectangular signal with (a) RC, (b) G_2.

where,

$$I_w^H = P_{\text{in}}^H G_s,\ I_w^L = \left(\left(P_{\text{in}}^H/\text{ER}\right)G_0 + P_{\text{in}}^H G_0 F\right),$$
$$\sigma_L^w = \sqrt{\left(P_{\text{ASE}}^0\right)^2 + 2P_{\text{ASE}}^0\left(P_{\text{in}}^H/\text{ER}\right)G_0 + 2P_{\text{ASE}}^0 P_{\text{in}}^H G_0 F},$$
$$\sigma_H^w = \sqrt{\left(P_{\text{ASE}}^s\right)^2 + 2P_{\text{ASE}}^s P_{\text{in}}^H G_s} \tag{27}$$

and,

$$Q_b = \frac{I_b^H - I_b^L}{\sigma_L^b + \sigma_H^b}$$

where,

$$I_b^H = P_{\text{in}}^H G_0,\ I_b^L = \left(\left(P_{\text{in}}^H/\text{ER}\right)G_s + P_{\text{in}}^H G_s F\right), \sigma_L^b = P_{\text{ASE}}^s,\ \sigma_H^b = \sqrt{\left(P_{\text{ASE}}^0\right)^2 + 2P_{\text{ASE}}^0 P_{\text{in}}^H G_0} \tag{28}$$

As an example, we consider an SOA with a small-signal gain of 17.2 dB ($P_{\text{in}}^{\text{sat}}$ = −13.8 dBm) at the operating wavelength of 1555 nm. The analytical results while applying different EL filters are depicted in **Figure 14**. It is clear that the filter impact is more intense when the signal is affected by SOA nonlinearities. Therefore, in this operation condition, an optimal filter bandwidth may be very different from the standard matched one.

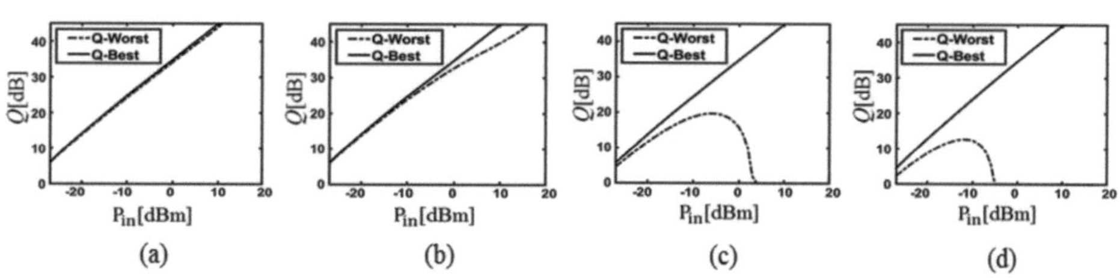

Figure 14.
Quality factor as a function of input power for 1555 nm signal, (a) without EL filter, (b) with RC filter, (c) with filter G_1, and (d) with filter G_2.

The model including different filters (RC, G_1, and G_2) have been validated numerically for an SOA input power large enough to assure a nonlinear regime (-4.8 dBm). We considered the effect of filters on signal quality by making a comparison of Q_w at a specific input power, –4.8 dBm, for each filter. The value of Q_w for the received signal without applying a filter is 26.4dB while it is reduced to 25.8 dB by applying the RC filter. It changes to 17.1, and 1 dB by applying G_1, G_2 filters, respectively. Such a huge change in Q_w highlight the importance of filter shape and parameters and demonstrate that matched filter used in the linear case may be not optimal. The analytically obtained results are in good agreement with the numerically simulated results (**Figures 12** and **13**).

As it is shown in **Figure 14a**, in the absence of an electrical filter, the signal quality at the high input power is high, even though the signal pattern in this area is affected by SOA nonlinearities. **Figure 14b** shows that the RC filter causes to reduce the Q_w at the high input powers, although RC induced lower effects than the other mentioned filters on signal quality. Referring to **Figure 14c**, applying G_1 on the received signal, leads to a decrease in the quality of the signal at the NL area of operation. This behavior is a consequence of the filter effects on the distorted signal. On the high input power side, the signal distortion occurs due to the patterning, however, the electrical filter induced the ISI between received symbols of the distorted signal. In the absence of the filter, the patterning does not cause to reduce the signal quality and there is no ISI between received symbols. Applying G_2 results in more severe ISI at high input powers and beyond a specific input power, the Q-factor becomes unmeasurable (**Figure 14d**). Since the value of the Super Gaussian filter on bits zero (F_{G2}) was found higher than the value of the Gaussian filter (F_{G1}), a higher ISI with G_2 was expected.

In conclusion, the electrical filter can be appropriate for optimizing performance in the linear operation to reduce the additive noise, however, for the nonlinear operation, it is inappropriate since it leads to a decrease in the Q-factor due to the patterning and ISI effects. Thus, system optimization must consider receiver filter shape and bandwidth as a function of SOA operation.

Finally, the performance of the G_1 on the amplified received signals, with various input power and different ERs, has been considered. In comparison with **Figure 11**, the effect of the filter on Q_w is more intense (**Figure 15b** while, as was expected from the analytical model, there is no significant quality degradation inQ_b. Hence, applying

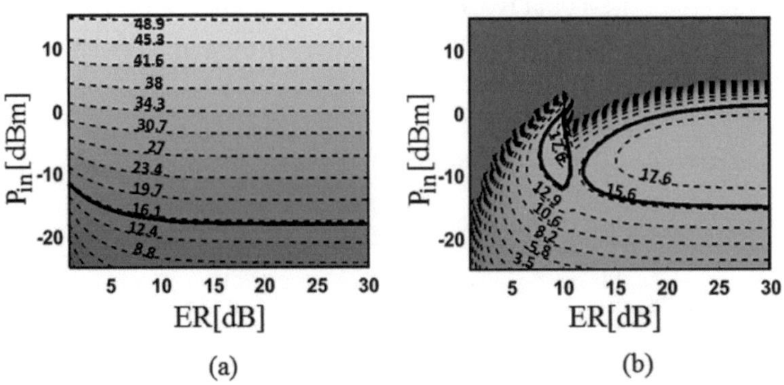

Figure 15.
First-order Gaussian filter effects on output signals of SOA with gain 15.3 dBm. Lines separate the boundary with different quality factors in the (a) best (Q_b), and (b) worst (Q_w) cases.

the G_1 electrical filter on the receiver side leads to degrading the signal quality on the high input power side due to the patterning and severe ISI (**Figure 15b**).

4. Experimental setup and measurement results

The analytical model is validated through experiments. For that purpose, SOA gain was characterized as a function of input power different bias currents (from 200 to 70 mA). Each bias current provides a specific G_0. **Figure 16** shows the experimentally measured gain as a function of input power for a wavelength of 1555 nm for a driving current of 200 mA (92 mA). Simulation is included for comparison using the following SOA parameters: 1800 μm-length and 0.38 μm-width active region, G_0 = 30.9 dB (G_0 = 17.2 dB) at the 1555 nm wavelength, confinement factor of 0.4, and the transparency carrier density is $3.07 \times 10^{18} cm^{-3}$.

The experimental setup for model validation is depicted in **Figure 17**. A 10 Gb/s NRZ signal was generated by exploiting a Mach-Zehnder modulator (MZM) fed with a continuous wave laser (CW) at 1555 nm and driven by a bit pattern generator (BPG) with a 2^{11}-1 PRBS. The input power into the SOA is adjusted through a variable optical attenuator (VOA). The MZM bias has been adjusted to keep a constant extinction ratio of 13 dB for the generated rectangular signal. This tuning is performed to emulate the ER effects on the signal performance. The signal input power and extinction ratio had been measured before launching into the SOA. The amplifier is biased at 70 mA, which corresponds to 15.3 dB gain for 1555 nm wavelength with saturation input power of –13 dBm. A 0.7 nm optical band-pass filter (OBPF) was used to mitigate the emitted ASE noise for eye-diagram evaluation and Q-factor. The received amplified signal is detected utilizing a 40 GHz photo receiver followed by a 45 GHz real-time oscilloscope and a 10 Gb/s error detector. To investigate the ER effects, in both linear and NL operation regimes, the measurements have been done on the signal with different powers ranging between –20 dBm and +3 dBm, fed into the SOA. **Figure 17** reports the eye diagram of the received signal in back-to-back and the eye diagram of the amplified signal, with +3 dBm injected power into the SOA. Although

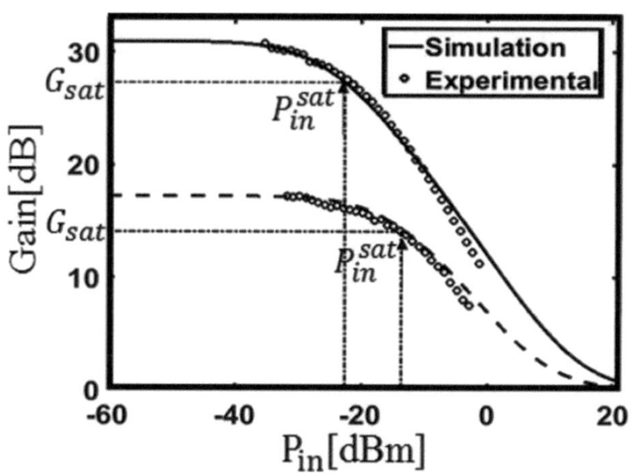

Figure 16.
Gain as a function of input power at I_{dc} = 200 mA (I_{dc} = 92 mA). Simulation (solid line) and measured (dots). The G_o and the P_{in}^{sat} are 30.9 dB (17.2 dB) and –24 dBm (–14 dBm), respectively.

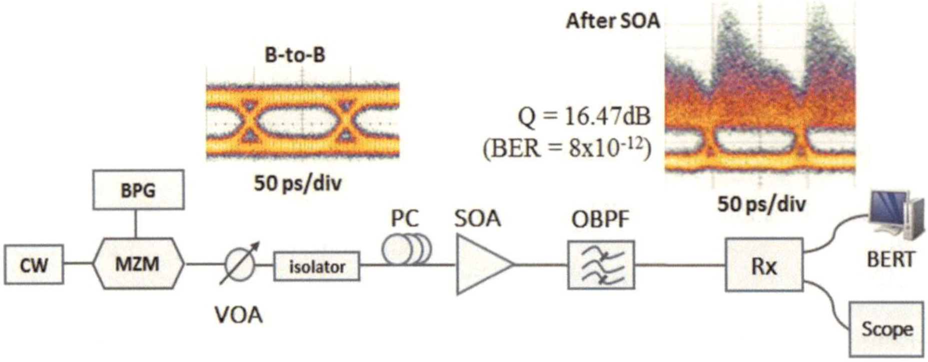

Figure 17.
Experimental setup for BER measurements of the amplified signal, while the input power is varied, the ER of the input signal to SOA is tuned to 13 dB (and 6 dB). The eye diagrams of the received signals are shown for the back-to-back (left) and after the SOA (right) with the input ER of 13 dB at 3 dBm input power. SOA drives at the bias current of 70 mA, which corresponds to a 15.3 dB gain for 1555 nm wavelength. BPG: bit pattern generator, OBPF: optical band-pass filter, PC: polarization controller, VOA: variable optical attenuator.

such a high input power drives the SOA into the deep saturation, the result shows a clear eye-opening with the quality factor of 16.4 dB (corresponding to a BER of 10^{-12}), which is in good agreement with the analytically obtained result. The analytically obtained signal quality as a function of input power for a constant extinction ratio of 13 dB is depicted in **Figure 18a** together with the experimental measurement results. The analytically Q-factor for the signal with the input power of –11 dBm is 20 dB while in practice obtaining this value of Q-factor requires a device with very high sensitivity. Therefore, we considered the maximum sensitivity of the available device, $Q = 16.9$ dB, which corresponds to a BER of 10^{-12} (dashed line in **Figure 18**) as the target for all obtained analytical quality factors which exceed 16.9 dB.

To evaluate the model for low ER effects on SOA operation, we also performed the BER measurements for the signal transmission with ER = 6 dB and various powers at the SOA input. Experimental BER results are depicted in **Figure 18b** together with

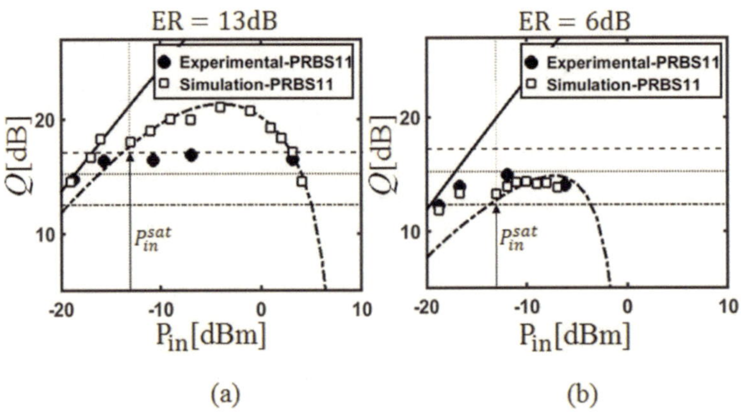

Figure 18.
Quality factor as a function of input power for 1555 nm signal with, (a) ER=13 dB, (b) ER=6 dB. The analytical Q_b and Q_w are shown with the solid lines and dashed curves, respectively. Simulation and experimentally obtained results are marked. The dash-dotted lines show the Q of 12.6 dB (BER = 10^{-5}), the dotted lines show the Q of 15.6 dB (BER = 10^{-9}), the dashed lines show the Q of 16.9 dB (BER = 10^{-12}).

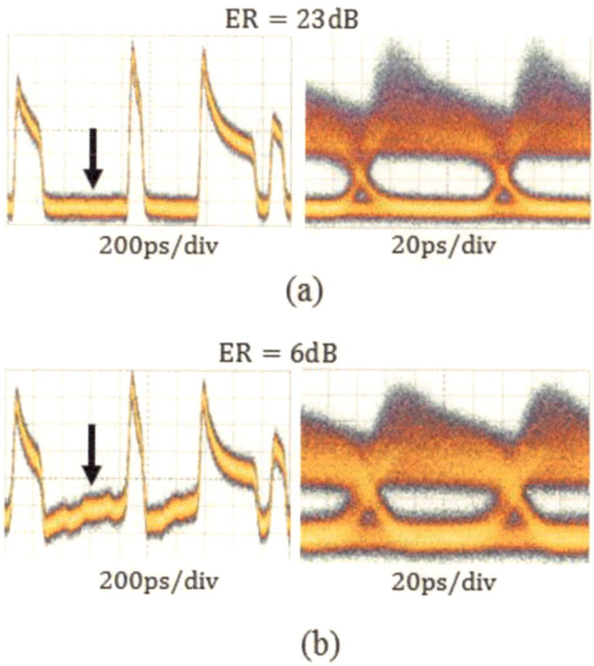

Figure 19.
Eye diagrams of received amplified signal with SOA input power of −4.8 dBm and (a) ER = 23 dB, (b) ER = 6 dB, at a bias current of 70 mA, with the pattern of the corresponding signals (left side). The arrows marked the ER effect on zero level.

both analytical and simulated results. The analytical curves and simulation results are in good agreement with experimentally obtained results.

We made a comparison between a data pattern with a low and high extinction ratio. In **Figure 19**, the eye diagrams of amplified signals with −4.8 dBm input power, ER = 6 and 23 dB, are reported along with the corresponding data patterns. From the signal's pattern, it is evident that low ER causes a severe patterning effect due to the distortion of zero level and a consequence eye degradation. While at high ER, there will be no ER effect on the zero level, and the patterning effect is only due to nonlinear effects.

Based on the experimental results, we can interpret the behavior of the SOA quality factor due to our analytical model as follow. We estimated a boundary condition for the SOA quality factor Q_b, and Q_w . When SOA operates in the linear regime, the signal quality factor is compatible with the analytical Q_b, and when the input power increases, the signal quality tends to the Q_w. At the input power above the SOA saturation input power (marked by an arrow in **Figure 18**), where SOA operates in the NL regime, the signal quality is consistent with the Q_w.

5. Discussion

The optimization of system performance requires proper knowledge of the device parameters, operating conditions, and system specifications when an SOA is used as a gain element. The availability of an analytical model capable of describing most of the dominant effects is beneficial in system design. Here, we proposed a method based on the analytical expressions for best and worst quality factors (Q_b and Q_w). Within this model, the effect of variable intrinsic and extrinsic SOA factors, such as small-signal gain value, signal extinction ratio, and filtering, on signal quality factors has been investigated. The model was developed for NRZ modulation formats. It can afterward be improved for advanced modulation formats such as QPSK or M-QAM. Exploiting our model, we investigated the performance of an SOA with moderated gain through the analysis of the Q-factor by considering the influence signal extinction ratio. It has been noticed that using low gain SOA for low *ER* signals gives a performance improvement. To evaluate the impact of the gain on the low *ER* signals, we performed the contour plot for the Q-factor of an amplified signal over a wide range of input power at a low range of *ER*. SOA was driven with different currents (70 and 37 mA), which results in various small-signal gain at 1555 nm. The plot in **Figure 20a** shows the obtained Q_w at the small-signal gain of 15.3 dB and the results of 8.2 dB gain ($P_{in}^{sat} = -8.3$ dBm) are depicted in **Figure 20b**. As represented in **Figure 20**, using SOA with the high gain for the low input ER causes degradation due to intense patterning.

The following equations provide us with a guideline to estimate the Q-factor boundary conditions, with a different range of extinction ratios and a large range of SOA input power. This tool also leads us to choose a proper electrical filter at the receiver side:

$$Q_w = \frac{I_w^H - I_w^L}{\sigma_L^w + \sigma_H^w}$$

where

$$\begin{aligned} I_w^H &= P_{\text{in}}^H G_s \\ I_w^L &= \left(\left(P_{\text{in}}^H/\text{ER}\right)G_0 + P_{\text{in}}^H G_0 F\right) \\ \sigma_L^w &= \sqrt{\left(P_{\text{ASE}}^0\right)^2 + 2P_{\text{ASE}}^0\left(P_{\text{in}}^H/\text{ER}\right)G_0 + 2P_{\text{ASE}}^0 P_{\text{in}}^H G_0 F} \\ \sigma_H^w &= \sqrt{\left(P_{\text{ASE}}^s\right)^2 + 2P_{\text{ASE}}^s P_{\text{in}}^H G_s} \end{aligned} \tag{29}$$

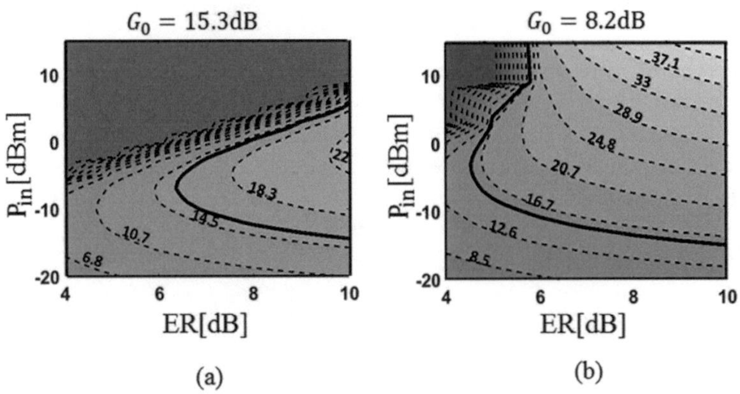

Figure 20.
Signal performance dependence on the input power and ER for an SOA with a small-signal gain of (a) G_o = 15.3 dB, and (b) G_o = 8.2 dB. Lines separate the boundary with different quality factors (Q_w). The solid lines mark the boundary with error-free amplification of a data stream where the Q = 15.6 dB (BER of 10^{-9}).

and,

$$Q_b = \frac{I_b^H - I_b^L}{\sigma_L^b + \sigma_H^b}$$

where

$$\begin{aligned} I_b^H &= P_{\text{in}}^H G_0 \\ I_b^L &= \left(\left(P_{\text{in}}^H / \text{ER} \right) G_s + P_{\text{in}}^H G_s F \right) \\ \sigma_L^b &= P_{\text{ASE}}^s \\ \sigma_H^b &= \sqrt{\left(P_{\text{ASE}}^0 \right)^2 + 2 P_{\text{ASE}}^0 P_{\text{in}}^H G_0} \end{aligned} \tag{30}$$

5.1 Lossless PICs

We propose to use this tool as guidelines to make a design for lossless photonic integrated circuits. There are plenty of research activities on the design and performance of PICs in the form of functional devices for standard networks and transmission. The standard networks require advanced PICs with the lowest loss. Several investigations have been done on providing new designs and technologies for the fabrication of low-loss PICs. However, using SOA in PICs, as a loss compensator can be an effective way of mitigating PICs losses. We propose the design guidelines for the SOA as an on-chip loss compensator, which provides a high-quality amplified signal to optimize the PIC performance. Exploiting the analytical model given in this manuscript, accompanied by numerical analysis of a multi-section cavity model we proposed in [41], can lead us to realize the optimum design of an SOA on-chip. For optimal SOA model design to incorporate the flip-chip bonding technique constraints in the hybrid integrated SOA–SOI chip, we proposed to exploit the multi-section analysis [41]. Our analytical model provides us with a guideline to easily design lossless PICs, using an SOA for loss compensation. Within this model, several effects, including reduced input extinction ratio, on SOA performance are considered. In addition, it was shown how these effects, can be compensated only by changing the SOA gain. Based on our analytical model, it is suitable to use a low gain SOA to ensure that signals are amplified with high quality, no matter the input signals extinction ratio is high or low. To illustrate that, we considered a PIC with an estimated total loss of about 10–12 dB. The estimated boundary conditions of quality factors [Eqs. (29) and (30)] guide us to choose an SOA with the proper gain for compensating the presented loss. Considering –9 dBm input signal, at the wavelength of 1555 nm with an extinction ratio of 5 dB, drives into the SOA.

As depicted in **Figure 21**, using 15 dB gain compensates 100% of loss at the cost of reducing Q-factor while using 10 dB gain, 70% of loss will be compensated but ensures an error-free operation. The reason behind the Q-factor reduction at this power, with 15 dB gain, is the OSNR degradation due to accumulated ASE noise emanating from the SOA. In our SOA model, the ASE noise is modeled as white Gaussian noise. Accordingly, we considered the affection of the accumulated noise from SOA on the Q-factor as noise variances. The effect of ASE in lower gain (10 dB) is smaller than that in higher gain (15 dB) at the input power of –9 dBm for 5 dB of ER. We should note that the –9 dBm input power is equal to the saturation input power of

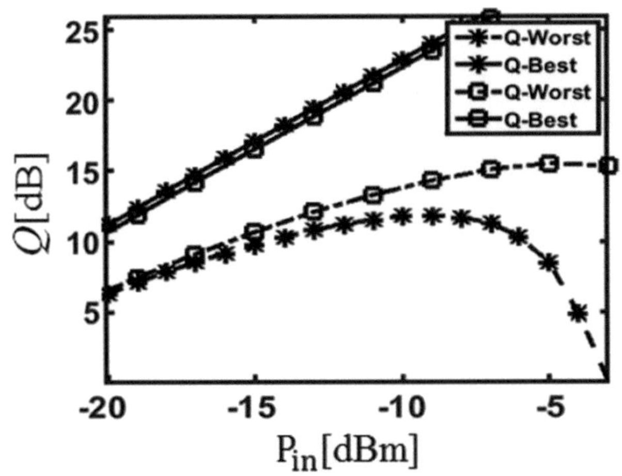

Figure 21.
Quality factor boundary of the SOA with G_o = 15.3 dB (marked by stars) and saturation input power of –13 dBm, and the SOA with G_o = 10 dB (marked by squares) and saturation input power of –9 dBm, for an input signal at 1555 nm with ER = 5dB.

the SOA with 10 dB gain, and it drives the SOA with 15 dB gain into the nonlinear operation area. However, the saturated gain of the SOA, with 15 dB gain, at this input power is higher than the saturated gain of the other SOA (with 10 dB gain), which causes a higher ASE and a reduction in Q-factor. Another reason for the Q-factor reduction is the impact of the gain value on the zero-level bits. Since the ER is low, the higher gain has a severe impact on zero-level bits, consequent the Q-factor reduces. Although, at low input power both gains show the same behavior, though using low gain SOA is more appropriate at the higher input power. The results show only 10 dB SOA is enough for compensating 10 dB loss for both linear and nonlinear operation areas.

5.2 Optimize SOA parameters

This tool also can be used as design guidelines to maximize the amplified signal Q-factor based on the SOA parameters. Designing an active region with a short length, large-cross section, and low confinement factor produces low G_0, high P_{in}^{sat}, and large gain bandwidth [47]. Based on the type of the active region, the confinement factor value would be different, such that the quantum dot layers and, in most cases, the quantum wells have confinement factors lower than bulk materials. The amount of confinement factor influences the saturation power so that for a specified active region cross-section, decreasing confinement factor leads to an increase in the saturation power. It is appropriate to mention that Γ is the fraction of the optical mode that overlaps with the active medium, therefore, a very low Γ causes the signal to expand broadly out of the active region and leaks into the surrounding regions. After adapting the device geometry and material, choosing a proper bias current is required for obtaining the desired gain. For fixed geometrical parameters by defining the value of bias current, we can control the gain value, peak wavelength, injected carrier density, and carrier lifetime. Carrier lifetime is inversely proportional to the injected bias current, therefore, decreasing injected current increases carrier lifetime, which leads to a decrease in the saturation power. Furthermore, by decreasing bias current and, accordingly, carrier density decreases G_0 and moves the peak to higher wavelengths.

It is worth mentioning, with a detailed perspective, we should consider that at very low currents, the stimulated emission (SE) dominates τ_c, and at very large currents, the Auger recombination and spectral hole burning (SHB) lead to the different behavior of τ_c. Where, for the reason of simplification, we approximated the carrier lifetime inversely proportional to the injected bias current. This simplification is concerning using only one rate equation to describe all carriers in the device [Eq. (31)] [40].

The boundary conditions of the quality factor show that the Q-factor behavior is related to the value of G_0 and n_{sp}. Note that the P_{ASE} is proportional to the n_{sp}. Therefore, in addition to choosing the proper gain, optimizing the population-inversion factor leads to optimizing Q-factor. As an example of using this study as a guideline for designing SOA, we considered the performance of an SOA with different active lengths but with the same peak gain and peak wavelength and the same saturation power. Achieving this, we start considering a device with a small active length and the peak gain of about 8 dB at 1555 nm, then the length increases, and the bias current changes accordingly to attain the same peak gain as the device with the smaller length. All other device parameters, including the confinement factor, remain constant. In such a manner, the carrier density, therefore, the population-inversion factor, changes by changing the bias current and active length.

Initially, we modeled SOA based on the experimentally obtained and fit parameters, using the numerical model described in [41]. Afterward, we changed the active region length and some other parameters to achieve the ≈ 8 dB gain. We modeled SOAs with different active region lengths (L) while trying to keep G_0 and P_{sat} constant. Somehow, at large L, the injected carriers (n) by the same bias current are not enough to provide the same gain as the small length SOA. To overcome this fact, we increase I_{dc} with the proper value, which leads to injecting enough carriers to attain the same gain. The injected carriers' value at unity quantum efficiency is given by [42].

$$n = \frac{I_{dc}\tau_c}{q\sigma_m L} \tag{31}$$

where q is the electron charge. The I_{dc} and L increased somehow to have lower n at larger L that corresponds to higher n_{sp} at larger L, where $n_{sp} = \frac{n}{n-n_{tr}}$ and n_{tr} is transparency carrier density. Observing Q-factor for different active region lengths, while the peak wavelength, G_0, and P_{sat} kept being constant, shows that the effects of SOA lengths on n_{sp} impact Q-factor (**Figure 22**). The parameters used in the simulations are presented in **Table 1**.

With regards to Eqs. (29) and (30), while the extinction ratio of the input signal is high and there is no electrical filter on the receiver side ($F = 0$), the Q-factor will be influenced by the SOA parameters and the input power. In this example, we considered Q-factor behavior performance in the linear operation regime at the input power of –15 dBm with the extinction ratio of 6 dB. The length effect of the active region on the Q-factor behavior is depicted in **Figure 22**.

When the length decreased, the peak gain was maintained constant by reducing the bias current simultaneously. Decreasing bias current decreases the population-inversion factor, which results in optimized Q-factor boundary conditions. Thereby a device with low G_0 and low n_{sp} with short active region length is expected to work with error-free amplification. It should be mention that decreasing the length increases the gain bandwidth, therefore, while the peak wavelength could remain

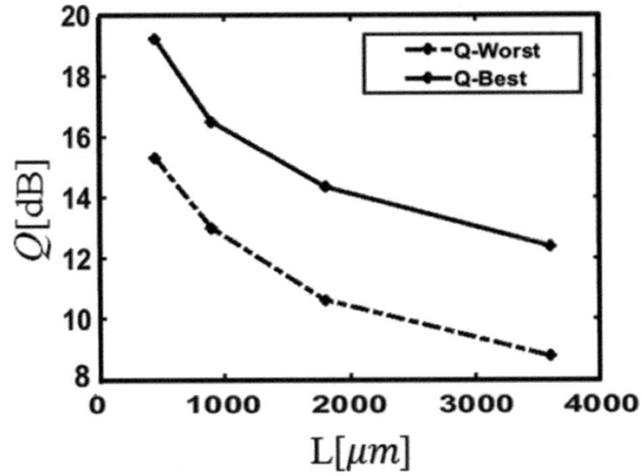

Figure 22.
Quality factor boundary of the linear operation area of an SOA with different active regime length, where its peak gain and the saturation power remains constant. The input signal power and extinction ratio are −15 dBm and 6 dB, respectively.

Symbol	Description	value	Unit
w	Width of active region	0.4	μm
d	Thickness of active region	0.1	μm
Γ	Confinement factor	0.4	—
n_{tr}	Transparency carrier density	3.08×10^{18}	cm^{-3}
α	Internal loss coefficient	3.5	cm^{-1}
a	Differential gin coefficient	1.7×10^{-16}	cm^2
$\bar{n}$	Average refractive index	3.45	-
τ_c	Free carrier life time	382	Ps
λ_p	Peak wavelength	1555	nm

Table 1.
SOA geometrical and material parameters were used in the simulation.

constant by reducing the bias current, the side wavelengths, far from the peak, might take different gain in different active region lengths.

6. Conclusion

We proposed an analytical model for estimating the performance of optical transmission systems including SOA in terms of Q-factor. The model allows to design high-performance SOAs based on system design constraints and requirements. In PIC with integrated SOAs, the model accompanied by the multi-section cavity model we proposed in [41] provides a tool for improving the design, limiting the number of fabrication runs and related costs. The model estimates the Q-factor based on SOA parameters (gain), signal properties (ER), and filter type at the receiver side. The effects of ASE noise and saturated gain compression are also included in the model. The model was validated through numerical simulation and experiments. The dependency of the optimal SOA gain on the signal extinction ratio has been investigated. This fact is a crucial aspect in optical interconnection applications that use low extinction ratio transmitters. It was shown at a low extinction ratio ($ER < 10$ dB), the fully recovered gain changes with input power and ER. Based on our model, we proposed the design of a lossless PIC, employing low gain SOAs for obtaining a high-quality amplified signal. It has been demonstrated that using low gain SOAs in PICs allow counteracting signal distortions, in both linear and nonlinear operation regime. In addition, we have studied the impact of different types of receiver filters on the performance of an amplified signal. It was shown that using a Gaussian filter leads to reduced signal quality due to patterning effects and ISI. Nevertheless, the impact of an RC filter is less intense, so confirming that the selection of a proper filter is a significant task while dealing with an amplified signal.

Acknowledgements

The authors thank Dr. Francesco Fresi for fruitful discussions and invaluable suggestions.

Conflict of interest

The authors declare no conflict of interest.

Author details

Pantea Nadimi Goki[1]*, Antonio Tufano[2], Fabio Cavaliere[3] and Luca Potì[4,5]

1 TeCIP Institute Scuola Superiore Sant'Anna, Pisa, Italy

2 Università Pegaso, Naples, Italy

3 Ericsson, Pisa, Italy

4 CNIT Photonic Networks and Technologies National Laboratory, Pisa, Italy

5 Università Mercatorum, Rome, Italy

*Address all correspondence to: pantea.nadimigoki@santannapisa.it

References

[1] Hulme JC, Doylend JK, Heck MJR, Peters JD, Davenport ML, Bovington JT, et al. Fully integrated hybrid silicon two dimensional beam scanner OSA. Optics Express. 2015;**23**:5861-5874

[2] Pérez D, Gasulla I, Crudgington L, Thomson DJ, Khokhar AZ, Li K, et al. Multipurpose silicon photonics signal processor core. Nature Communications. 2017;**8**:1-9

[3] Vlasov YA. Silicon CMOS-integrated Nano-photonics for computer and data communications beyond 100G Authorized licensed. IEEE Communications Magazine. 2012;**50**: S67-S72

[4] Nadimi GP, Imran M, Fresi F, Cavaliere F, Potì L. Lossless ROADM by exploiting low gain SOAs in fronthaul network. In: Proceedings of the IEEE International Conference on OptoElectronics and Communications/ Photonics in switching and computing (OECC/PSC); 7–11 July 2019; Fukuoka. Japan: IEEE; 2019. pp. TUG3-TUG3

[5] Matsumoto T, Kurahashi T, Konoike R, Suzuki K, Tanizawa K, Uetake A, et al. Hybrid-integration of SOA on silicon photonics platform based on flip-chip bonding. IEEE Journal of Lightwave Technology. 2019;**37**:307-313

[6] Spiekman L. Current commercial applications of semiconductor optical amplifiers. In: Proceedings of the IEEE International Conference on Transparent optical networks (ICTON); 6–10 July 2014. Graz. Austria: IEEE; 2014. p. Mo.C2.1

[7] Renaudier J et al. First 100 nm continuous-band WDM transmission system with 115Tb/s transport over 100km using novel ultra-wideband semiconductor optical amplifiers. In: Proceedings of the IEEE International Conference on Optical communication (ECOC); 17–21 September 2017; Gothenburg. IEEE: Sweden; 2018

[8] Cho PS, Achiam Y, Levy-Yurista G, Margalit M, Gross Y, Khurgin JB. Investigation of SOA nonlinearities on the amplification of DWDM channels with spectral efficiency up to 2.5 b/s/Hz. IEEE Photonics Technology. 2004;**16**: 918-920

[9] Assadihaghi A, Teimoori H, Millett R, Benhsaien A, Tolstikhin V, Hall T, et al. O-band semiconductor optical amplifier design for CWDM applications. In: Proceedings of the IEEE International Conference on Microsystems and nanoelectronics (MNRC); 15 October 2008. Ottawa, Ontario, Canada: IEEE; 2008. pp. 89-92

[10] Schmuck H, Bonk R, Poehlmann W, Haslach C, Kuebart W, Karnick D, et al. Demonstration of an SOA-assisted open metro-access infrastructure for heterogeneous services. Optics Express. 2014;**22**:737-748

[11] Bonk R. SOA for future PONs. In: Proceedings of the IEEE International Conference on optical fiber communication (OFC); 11–15 March 2018. San Diego, CA, USA: IEEE; 2018. pp. 1-3

[12] Ramírez-Cruz E, Rendón-Salgado I, Ceballos-Herrera DE, Gutiérrez-Castrejón R. Cross-Gain modulation mitigation in semiconductor optical pre-amplifiers using a new synthetic filter for NRZ-based WDM transmission systems. In: Proceedings of the IEEE International Conference on Transparent optical networks (ICTON); 1–5 July 2018. Bucharest, Romania: IEEE; 2018. p. Th.B4.4

[13] Leuthold J, Ryf R, Maywar DN, Cabot S, Jaques J, Patel SS. Nonblocking all-optical cross connect based on regenerative all optical wavelength converter in a transparent demonstration over 42 nodes and 16,800 km. IEEE Journal of Lightwave Technology. 2003;**21**:2863-2870

[14] Stubkjaer KE. Semiconductor optical amplifier-based all-optical gates for high-speed optical processing. IEEE Journal of Selected Topics in Quantum Electronics. 2000;**6**:1428-1435

[15] Tangdiongga E, Liu Y, de Waardt H, Khoe GD, Koonen AMJ, Dorren HJS, et al. All-optical demultiplexing of 640to40Gbits/s using filtered chirp of a semiconductor optical amplifier. Optics Letters. 2007;**32**:835-837

[16] Matsumoto T, Kurahashi T, Konoike R, Tanizawa K, Suzuki K, Uetake A, et al. In-line optical amplification for silicon photonics platform by flip-chip bonded InP-SOAs. In: Proceedings of the IEEE International Conference on Fiber communication (OSA-OFC);11–15 March 2018. San Diego, CAUSA: IEEE; 2018. pp. 1-3

[17] Lazarou I, Stamatiadis C, Schrenk B, Stampoulidis L, Zimmermann L, Voigt K, et al. Colorless ONU with discolored source and hybrid SOI integrated wavelength converter. IEEE Journal of Photonics Technology. 2012; **24**:386-388

[18] Gasse KV, Wang R, Roelkens G. 27 dB gain III-V-on-silicon semiconductor optical amplifier with > 17 dBm output power. Optics Letters. 2019;**27**:293-302

[19] Moscoso-Mártir A et al. 8-channel WDM silicon photonics transceiver with SOA and semiconductor mode-locked laser. Optics Letters. 2018;**26**: 25446-25459

[20] Ma H, Chen S, Yi X, Zhu G. 1.55 μm spot-size converter integrated polarization-insensitive quantum-well semiconductor optical amplifier with tensile-strained barriers. IOP Semiconductor Science and Technology. 2004;**19**:846

[21] Qing G, Yesgaiahu F. Semiconductor Nanolasers. 1st ed. Cambridge, United Kingdom: Cambridge University Press; 2017. pp. 234-236

[22] Caillaud C, Chanclou P, Blache F, Angelini P, Duval B, Charbonnier P, et al. Integrated SOA-PIN detector for high-speed short reach applications. IEEE Journal of Lightwave Technology. 2015;**33**:1596-1601

[23] Caillaud C, Glastre G, Lelarge F, Brenot R, Bellini S, Paret JF, et al. Monolithic integration of a semiconductor optical amplifier and a high-speed photodiode with low polarization dependence loss. IEEE Journal of Photonics Technology. 2012; **24**:897-899

[24] Volet N, Spott A, Stanton EJ, Davenport ML, Chang L, Peters JD, et al. Semiconductor optical amplifiers at 2.0-μm wavelength on silicon. Journal of Laser Photonics Reviews. 2017;**11**:1-7

[25] Viheriala J, Tuorila H, Zia N, Cherchi M, Aalto T, Guina M. 1.3 μm U-bend traveling wave SOA devices for high efficiency coupling to silicon photonics. In: Proceedings of the SPIE-OPTO 10923 International Conference on Silicon Photonics XIV (SPIE-OPTO 10923); 4 March 2019. San Francisco, Cal, USA: SPIE-OPTO; 2019. p. 10923

[26] Bonk R. Linear SOA and network applications. In: Leuthold J, Freude W, Koos C, editors. Linear and Nonlinear Semiconductor Optical Amplifiers for Next-Generation Optical Networks.

Karlsruhe, Germany: KIT Scientific Publishing; 2013, ch.4, pp.117-127 and ch.2 pp.44

[27] Guan Y, Wang R. Demonstration of an optical switch based on SOA-MZI operation at 10 Gbit/s. In: Proceedings of the 2013 the International Conference on Artificial Intelligence and Software Engendering (Atlantis Press – ICAISE); July 2013; Zheng Zhou. China: Atlantis Press – ICAISE; 2013. pp. 239-241

[28] Reid DA, Clarke AM, Yang X, Maher R, Webb RP, Manning RJ, et al. Characterization of a turbo-switch SOA wavelength converter using spectrographic pulse measurement. IEEE Journal of Selected Topics in Quantum Electronics. 2008;**14**:841-848

[29] Kaur G, Kaler RS, Kwatra N. Performance investigation of semiconductor optical amplifier and Raman amplifier as an optical sensors Optoelectron. Advanced Materials. 2015; **9**:1110-1113

[30] Palagarla BB, Das N, Razaghi M. Analysis of nonlinear pulse propagation and wave-mixing characteristics in SOAs. In: Proceedings of the IEEE International Conference on Sensing technology (IEEE- ICST); 4–6 December 2017. IEEE: Sydney, NSW, Australia; 2018

[31] Armstrong I, Andonovic I, Kelly AE. Semiconductor optical amplifiers: Performance and applications in optical packet switching OSA. Journal of Optical Communications and Networking. 2004; **3**:882-897

[32] Spencer DT et al. An optical-frequency synthesizer using integrated photonics. Nature. 2018;**557**:81-86

[33] Hulme J, Kenndy M, Chao R, Liang L, Komljenovic T, Shi J, et al. Fully integrated microwave frequency synthesizer on heterogeneous silicon-III/V. Optics Express. 2017;**25**:2422-2431

[34] Komljenovic T, Liang L, Chao R, Hulme J, Srinivasan S, Davenport M, et al. Widely-tunable ring-resonator semiconductor lasers MDPI. Applied Sciences. 2017;**7**:1-21

[35] Johnson B, Atia W, Woo S, Melendez C, Kuznetsov M, Ford T, et al. Tunable 1060nm VCSEL copackaged with pump and SOA for OCT and LiDAR. In: Proceedings of the SPIE 10867 International Conference on Optical Coherence Tomography and Coherence Domain Optical Methods in Biomedicine XXIII (SPIE-BiOs10867); 22 February 2019. San Francisco, Cal, USA: SPIE-BiOs; 2019. p. 10867

[36] Säckinger E. Optical Transmitters. In: Broadband Circuits for Optical Fiber Communication. 1th ed. USA: John Wiley & Sons. Inc; 2005. pp. 233-257

[37] Lu H, Liaw J, Ji Y, Tsai W, Lee Y. A-10Gbit/s light wave transport system based on VCSEL and SOA with external light injection technique. IEICE Electronics Express. 2004;**1**:228-232

[38] Park H, Bowers JE. A hybrid AlGaInAs-silicon evanescent amplifier. IEEE Photonics Technology Letters. 2007;**19**:230-232

[39] Hulme JC, Doylend JK, Heck MJR, Peters JD, Davenport ML, Bovington JT, et al. Fully integrated hybrid silicon two dimensional beam scanner. Optics Express. 2015;**23**:5861-5874

[40] Bonk R, Vallaitis T, Guetlein J, Meuer C, Schmeckebier H, Bimberg D, et al. The input power dynamic range of a semiconductor optical amplifier and its relevance for access network

applications. Journal of IEEE Photonics. 2011;**3**:1039-1053

[41] Nadimi GP, Imran M, Porzi C, Toccafondo V, Fresi F, Cavaliere F, et al. Lossless WDM PON photonic integrated receivers including SOAs. MDPI Journal of Applied Science. 2016;**9**:1-14

[42] Agrawal GP. Optical Receivers. In: Fiber-Optic Communication Systems. 3th ed. New York, NY, USA: John Wiley & Sons; 2002. pp. 133-177

[43] Saleh AAM, Habbab IMI. Effects of semiconductor-optical-amplifier nonlinearity on the performance of high-speed intensity-modulation lightwave systems. IEEE Transactions on Communications. 1990;**38**:839-846

[44] Dupertuis MA, Pleumeekers JL, Hessler TP, Selbmann PE, Deveaud B, Dagens B, et al. Extremely fast high-gain and low-current SOA by optical speed-Up at transparency. IEEE Photonics Technology Letters. 2000;**12**:1453-1455

[45] Hale D. Recursive Gaussian filters. Center for Wave Phenomena. USA: Colorado School of Mines; 2006. p. CWP-546

[46] Roy TK, Pervej MF, Morshed M. Performance comparison of three optimized alternative pulse shaping filters with the Raised Cosine filter for wireless applications. In: Proceedings of the IEEE International Conference on Computer and Information engineering (IEEE- ICCIE); 26–27 November 2015; Rajshahi. Bangladesh: IEEE; 2016. pp. 9-12

[47] Mahony MJO. Semiconductor laser optical amplifiers for use in future fiber systems. IEEE Journal of Lightwave Technology. 1988;**6**:531-544

Chapter 3

Power Reduction Using Efficient Way of Tri-State Buffer Connection

Maan Hameed

Abstract

Clock gating is a very important technique for decreasing wasted power in digital design. One of the approaches to obtain dissipated power is an intention by the way of masking the clock pulse that is going to the unused part of the design. In this research, a comparative evaluation of current clock gating techniques on synchronous digital design changed into provided. In the new suggested design, the gated clock technology circuit is a use of tri-state buffer and gated clock. The new submodule was created by the connection of two tri-state logic used as switched to control to the design. The new suggested technique was saving more power and area. The suggested sub-module was achieved by using ASIC design methodologies. In order to implement Huffman modules, the architecture of the proposed module has been generated using Verilog HDL language. In addition, it is proved using Modalism-Altera 10.3c (Quartus II 14.1) tools. By using the tri-state technique, dynamic power and total power are decreased. The suggested technique will decrease the hardware complexity.

Keywords: tri state buffer, VLSI, Xilinx, glitches, hazards

1. Introduction

Incorporated Lap (ICS) terminus, normally indicate to surely chips or microchips, became accompanied with the aid of the want to test these designs. Small weighing device integrating (SSI) design, with tens of digital transistor in the early time of the 60s, and Medium Scale Leaf of size integration (MSI) system of policies, with a large quantity of digital transistor inside the overdue time 1960s, had been comparatively clean to test [1]. Moreover, within the 1970s, massive-scale integration (LSI) innovation, with a big range of transistors, generate some difficulties while experimenting with these designs. In the early time of 80s, a very-large-scale integration (VLSI) sample with a big range of transistors was explained. Enhance the caliber in VLSI designs have done in designs with hundreds of millions of transistors [2]. One of the primary dissipated ethical pressure exponents reduced by the clock gating approach, in calculating electronics ware inside the ordinary machine's gated clock. Then, decrease by 30–70% of the entire dynamic electricity dissipation. Similarly, lowering the general gadget strength to 15-xx% of grouping somersault fall apart statistics pushed the clock gating approach [3]. A gated clock may be a very crucial way of lowering the clock signal. Essentially, while clocked the common-sense social unit, it is relying upon the sequential parameter that receives the clock signal, one after the other, they may transfer within the subsequent HZ whether its miles demanded or now not. The facts driven clock gating designs figuring out allowing clock indicators are manually brought for every FF as a part of the proposed design methodology [4]. With the aid of the usage of the clock gating method, the fair sports clock sign is and with explicitly predefined allowing sign. Clock gating is appointed at all ground of electrical circuit, good judgment design, block layout, and gate [5]. Sundry ways to pick gain of this approach are defined, with all of them based on unique heuristic application in a looking to increase clock gating opportunities. With the quick ontogeny in gadget of regulations complexity. Laptop-aided layout (cad) tool supporting device stage-ironware-description has often been realized. Additionally, grow gadget output, every contraption wants the body of work of the long mountain variety of computerized synthesis algorithms, from register-switch-degree (RTL) right down to the netlist and gate-level. Sadly, such automation leads to a massive wide variety of unused toggling. Consequently, developing the dissipated clock at flip-flops as defined in this studies chapter [6].

2. Dynamic energy

Wasted power is created when the system is changing from a natural state to the next state or when switch ON. Changing power of one gate can be explained in the Eq. (1):

$$\mathrm{P}_{\mathrm{Dynamic}} = CLV^{2}{}_{DD}\mathrm{f} \tag{1}$$

where the parameter (f) represents the clock frequency, the parameter (CL) represents the switching capacitance, while the parameter (V_{DD}) represents the voltage source. Finally, decreasing the dynamic power consumption can be calculated by the way of decreasing one of the parameters in the Eq. (1) [7].

3. Reducing of dynamic power

Leakage power is the lowest amount of power dissipation in digital design, growing significantly every year with modern techniques. Dynamic power is the largest amount of power dissipation in the digital design, and still control all the power dissipated in the system. Therefore, the efficient techniques to drop off the dissipated dynamic power contain the size of the transistor, gated clock, multiple provision potential interconnect optimization, and dynamic control of supply voltage. Incorporating the above methods in the system of nanoscale system, the dynamic power consumption can be decreased significantly [8].

4. Clock gating

Clock gating is the efficient method used to decrease the power consumption in a synchronous system. In an ideal digital design like the general target microprocessor, part of the design is active only at any specific time. Therefore, by switching off the idle part of the design, the unused power dissipation can be saved. The most important approach to realize this goal by cutting a clock signal that goes to the unnecessary part of the design. This way stops unused clocking of the inputs to the idle system module [9].

5. Proposed technique

In this text, we will talk about the new device that will keep much area and power as proved in Ref. [10]. The brand-new signal named gated clock is shown inside the next

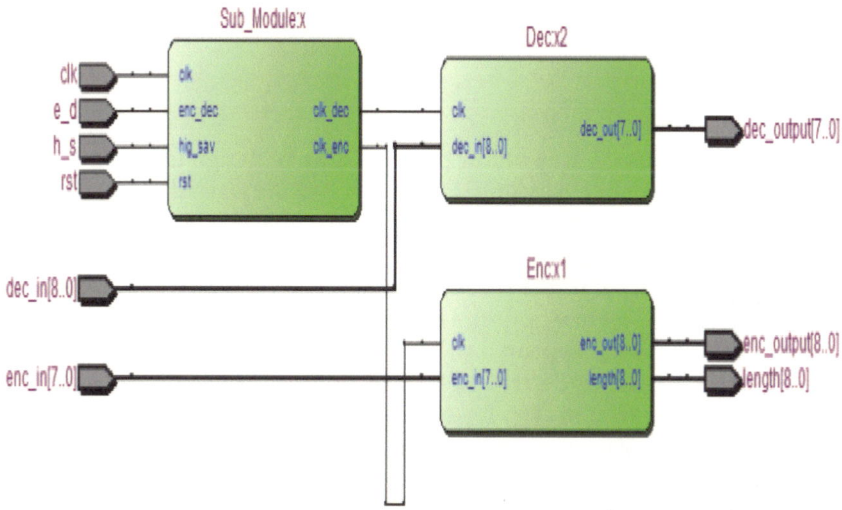

Figure 1.
RTL viewer for Huffman with PMC.

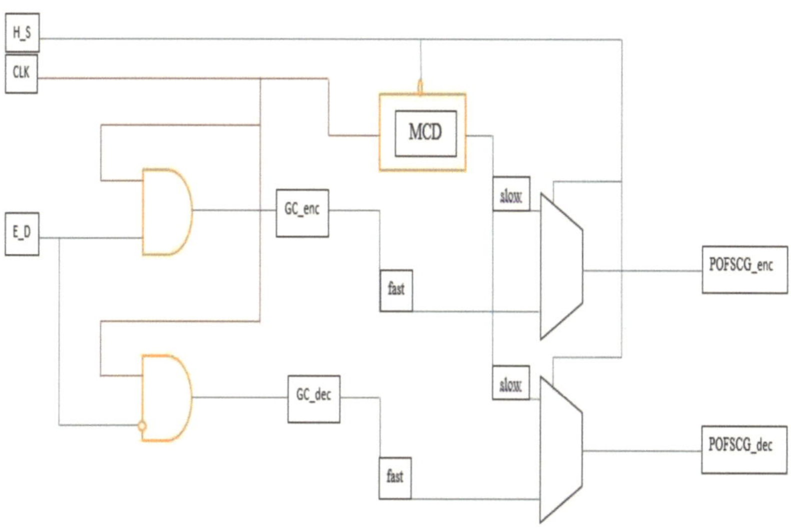

Figure 2.
Power management control (PMC) design.

determine later by means of including tri-nation connection and device of the logic gate is employed that is generated by way of the gathering of double-gated (and, or, and) with bubbled input sign sequent. On this conception, maintain baron in each case that even the truthful recreation designing is activated. The ascendance gadget clock is off, in summation when the specified design is clock off, and then additionally controlling twist's clock is off. This method appearing can maintain lots of energy by way of averting unused switching at the clock net. This clock gating is on at posedge and rancid at negedge stated clock electricity and dynamic power is maintaining on the jail time period of negedge clock. In the traditional state, the design works by applying a CLK signal controlled by enable signal (E-N).

Clear to understand the working of design in the original state from RTL, viewer for Huffman with PMC as shown in **Figures 1** and **2**.

6. Tri-state logic

This section explains the way of low power Huffman design using a tri-state buffer. By using tri-state buffer, significantly reduced the wasted power. Basically, tri-state logic is used as a switch and controlled by enable signal, named as Enable Signal (EN) as shown in **Figure 3**. In tri-state buffer, if en = 1, that led to appear the input data at the output of the tri-state buffer. In other word, encoder on and decoder off. But if en = 0, then opposite the process encoder off and decoder on. Also, input value would not appear at the output. Clearly, the process details are shown in **Figure 4**. In this work, tri-state buffer property is used to decrease the power dissipation of Huffman [11]. It performs various processes during the execution in modules block. At one particular time, one method was performed [12]. Even as, all different approaches (which are not select), consumes clock baron at the same time. This can leash dynamic power dissipation.

The problem is overcome with the usage of the tri-country buffer. Therefore, at the time one technique can carry out via Huffman, whilst all different operations are in tri-state high impedance z nation. In this example, it's miles feasible to put in force circuit with the assist of tri-kingdom good judgment. That results in carrying out the

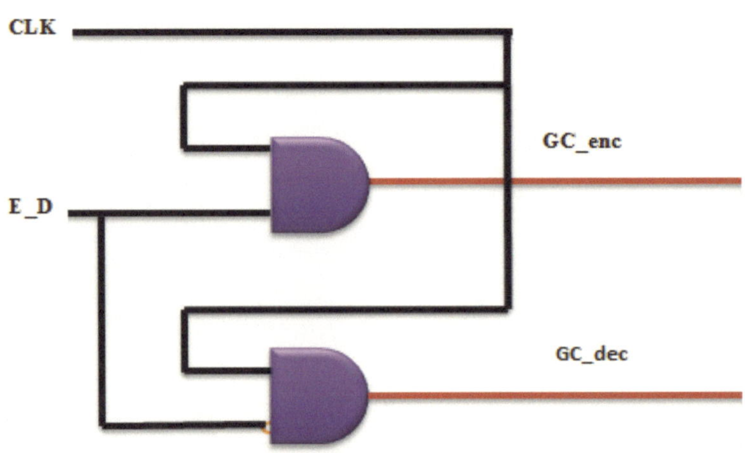

Figure 3.
Clock gating techniques.

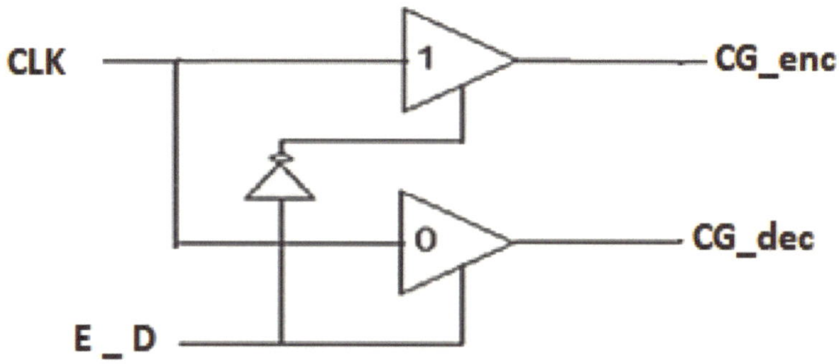

Figure 4.
Tri-state connection.

most effective one operation at one plus time, that is selection enter for selected the target method [6]. Moreover, the other operations are remaining in a high impedance Z state. In other word, these operations cannot access to output or connect to the output. The path of passing clock signals in different cases is presented in **Figures 5** and **6**, respectively. The way of execution using tri-state logic is shown in **Figure 4**. Where the implementation processes with tri-state as the way of the clock signal. If en = 1, then the clock signal will be appearing at the decoder output. But when the en signal is set to 0 as shown in **Figure 4**, clock input will be an appearance at the encoder output because an inverter makes en input 0 to input 1. So, it deeds as an out-of-doors circuited route [13]. This could make sure that Huffman modules cognitive manner, handiest one module is chosen at a time and carried out, while the relaxation of cognitive procedure are in tri-nation [14]. **Figure 5** shows Power Management Control (PMC) validation. While the full Huffman waveform validations is presented in **Figure 6**.

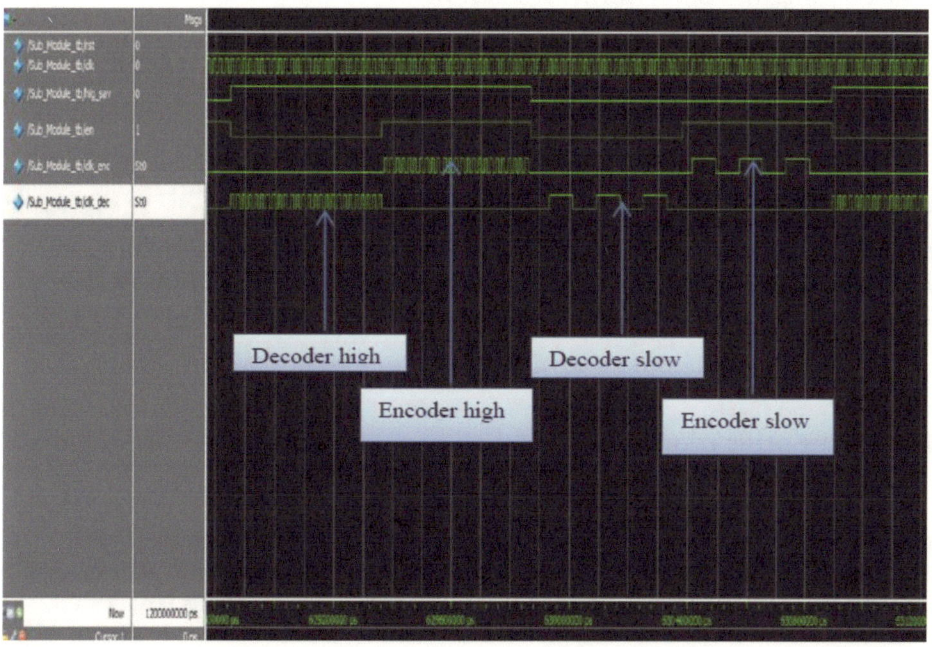

Figure 5.
Power management control (PMC) validation.

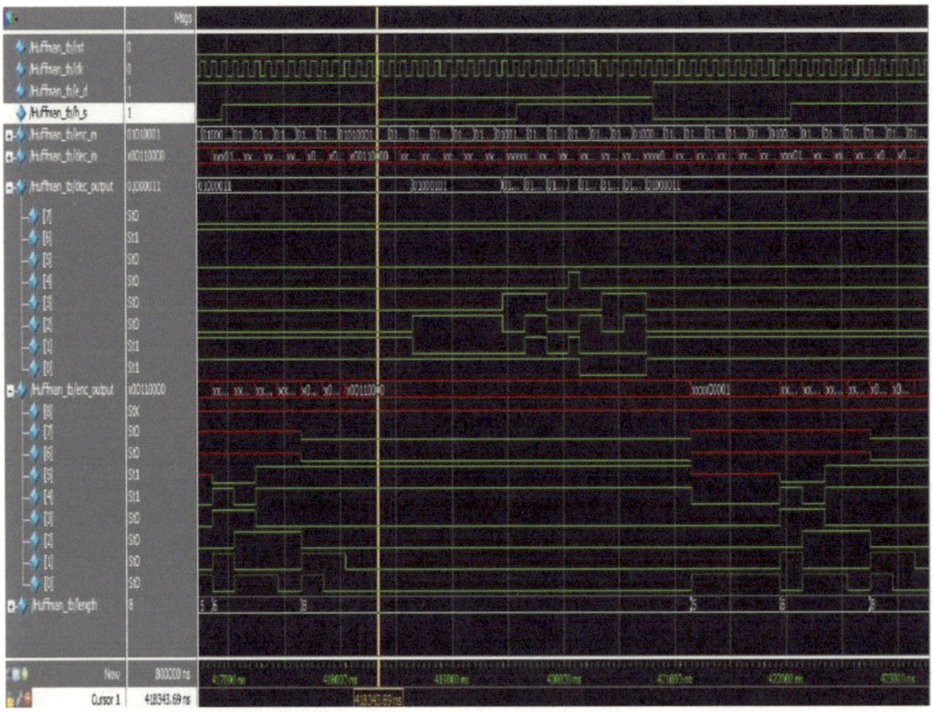

Figure 6.
Huffman waveform validations.

7. Validations and behavioral simulation

Quartus II 14.1 (Sixty-Four-Bit) web version is used to enforce Huffman. Furthermore, Modelsim-Altera 10.3c (Quartus II 14.1) is used to carry out behavioral simulation and validation [4]. Validation software is used for the behavioral simulation and tested the Huffman layout with exceptional values of entering and in all conditions. Huffman layout is behaving according to the specs. **Figures 5** and **6** show the behavior of the waveform simulation of Huffman with the new tri-state logic design, in addition to the behavioral of tri-state design. Clear to observe that, how the new circuit switches had one process ON and the other process OFF. **Figure 6** shows the validation of the input and output signals for Huffman with tri-state logic technique [15].

8. Resource utilization

In **Figure 7**, the resource utilization summary is shown. In this chapter, tri-state logic is used to control Huffman design. In this technique, encoder and decoder Huffman operations are implemented. Moreover, logic operations are achieved too. This will lead to less resource usage. After the compilation, synthesis is executed. Quartus II 14.1 (64-bit) internet version illustrated RTL schematic of Huffman. The RTL schematic is received, the device usage of the layout with tri-kingdom good judgment is proven under.

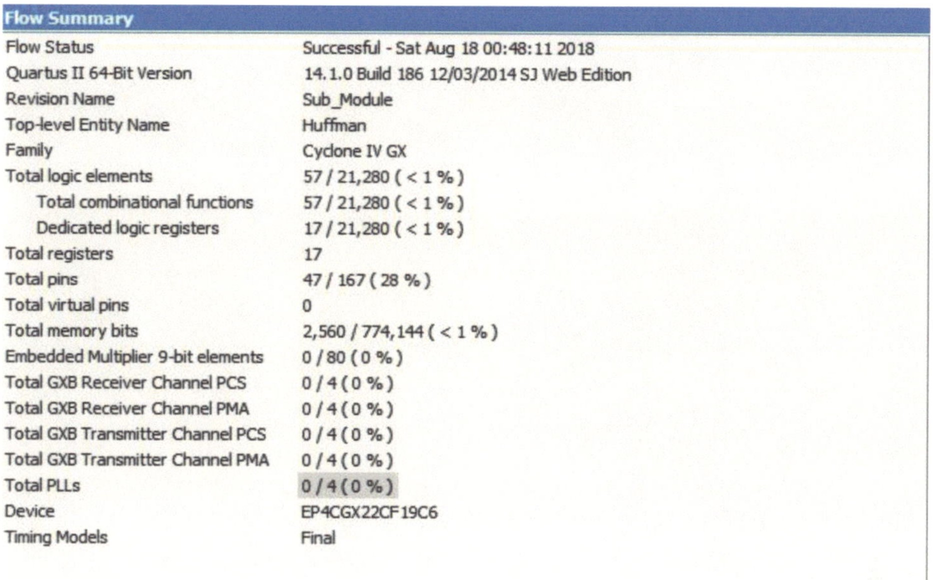

Flow Summary	
Flow Status	Successful - Sat Aug 18 00:48:11 2018
Quartus II 64-Bit Version	14.1.0 Build 186 12/03/2014 SJ Web Edition
Revision Name	Sub_Module
Top-level Entity Name	Huffman
Family	Cyclone IV GX
Total logic elements	57 / 21,280 (< 1 %)
Total combinational functions	57 / 21,280 (< 1 %)
Dedicated logic registers	17 / 21,280 (< 1 %)
Total registers	17
Total pins	47 / 167 (28 %)
Total virtual pins	0
Total memory bits	2,560 / 774,144 (< 1 %)
Embedded Multiplier 9-bit elements	0 / 80 (0 %)
Total GXB Receiver Channel PCS	0 / 4 (0 %)
Total GXB Receiver Channel PMA	0 / 4 (0 %)
Total GXB Transmitter Channel PCS	0 / 4 (0 %)
Total GXB Transmitter Channel PMA	0 / 4 (0 %)
Total PLLs	0 / 4 (0 %)
Device	EP4CGX22CF19C6
Timing Models	Final

Figure 7.
Flow summary results of Huffman.

9. Conclusion

Last of all, in this research the new suggested design will preserve more area and power. The most important aspect of this newspaper writer is to expand a new clock signal approach and improve the performance of the virtual design. The boom in dynamic essential multi-millionaire consumption makes the employer unreliable so that you could ascendency the dynamic transfer strength numerous techniques are studied to reduce it. A new tri-nation-based totally clock technique is proposed to reduce energy dissipation. Comparative analysis shows that the proposed approach affects the dynamic strength. All the analyses are done on Huffman design with process variation parameters. As concluded from the simulation waveform summarized in the figure above, the suggested technique deals with tri-state as a switch, not register. That's means switch used to pass the signal only. This fulfilled the purpose of this wok, reduction in power consumption. Finally, the performance evaluation of the various modules is carried out using Modalism-Altera 10.3c (Quartus II 14.1), which is used to perform behavioral simulation and validation, the circuits designed using tri-state logic showed a reduced power. As a future work, a reversible divider can also be designed and included in this design.

Author details

Maan Hameed
Ministry of Water Resources, State, Commission for Reservoirs and Dams, Iraq

*Address all correspondence to: maan_eng32@yahoo.com

References

[1] George L, Bangde P. Design and implementation of low power consumption 32-bit ALU using FPGA. International Journal for Research in Emerging Science and Technology. 2014;**1**(5):64-68

[2] Kathuria JA, Ayoub M, Khan M, Noor A. A review of clock gating technique. MIT International Journal of Electronics and Communication Engineering. 2011;**1**(2):106-114

[3] Paul BC, Agarwal A, Roy K. Low power design techniques for scaled technologies. Integration, the VLSI Journal. 2006;**39**(2):64-89

[4] Mohammed MH, Khmag A, Rokhani FZ, Ramli AB. VLSI implementation of Huffman design using FPGA with a comprehensive analysis of power restrictions. International Journal of Advanced Research in Computer Science and Software Engineering. 2015;**5**(6):49-54

[5] Chaudhary H, Goyal N, Sah N. Dynamic power reduction using clock gating: A review. IJECT International Journal of Electronics & Communication Technology. 2015;**6**(1):22-26

[6] Dokic B. A review on energy efficient CMOS digital logic. ETASR—Engineering Technology & Applied Science Research. 2013;**3**(6):552-561

[7] Kawa J. Low power and power management for CMOS—An EDA perspective. IEEE Transactions on Electron Devices. 2008;**55**(1): 186-196

[8] Raghavan N, Akella V, Bakshi S. Automatic insertion of gated clocks at register transfer level. In: International Conference on VLSI Design. 1999. pp. 48-54

[9] Soni DK, Hiradhar A. A review on existing clock gating. International Journal of Computer Science and Mobile Computing. 2015;**4**(3):371-382

[10] Hameed M, Mogheer HS, Razak I. Low power text compression for Huffman coding using Altera FPGA with power management controller. In: 2018 1st International Scientific Conference of Engineering Sciences IEEE—3rd Scientific Conference of Engineering Science (ISCES). 2018

[11] Hameed M, Khmag A, Rokhani FZ, Ramli AB. CMOS technology using clock gating techniques with tri-state buffer. Walailak Journal of Science and Technology. 2017;**14**(4)

[12] Hameed M, Rokhani FZ, Ramli AB. Low power approach for implementation of Huffman coding for high data compression. International Journal of Advances in Electronics and Computer Science. 2015;**2**(12)

[13] Karakehayov Z. Model-driven clock frequency scaling for control-dominated embedded systems. International Journal of Computing. 2014;**7**(2):100-107

[14] Hameed M, Khmag A, Rokhani FZ, Ramli AB. A new lossless method of Huffman coding for text data compression and decompression process with FPGA implementation. Journal of Engineering and Applied Sciences. 2016;**11**(3):402-406

[15] Hameed M, Mogheer HS, Mansour A. Power reduction using high speed with saving mode clock gating

technique. In: Paper Presented at the 2nd International Scientific Conference of Engineering Sciences (ISCES 2020). IOP Conference Series Materials Science and Engineering. Vol. 1076. Iraq: University of Diyala; 2021. p. 012055. DOI:10.1088/1757-899X/1076/1/012055

Chapter 4

Theory of Charge Transport in the Illuminated Semiconductor/Liquid Junctions

Peter Cendula

Abstract

The field of photoelectrochemical (PEC) cells for solar water splitting or CO_2 reduction has attracted intense attention of many research groups in last 15 years. Nevertheless, a cost-effective and efficient PEC cell for hydrogen production in the large scale was not yet discovered. The core functionality of the PEC cell is provided by the semiconductor/liquid junction, creating the electrostatic field to separate the photogenerated charges. This work aims to be a starting point for a newcomer in the field providing a compact knowledge about the charge transport and electrochemistry fundamentals in semiconductor/liquid junctions in the steady state. We describe charge transport within the semiconductor and electron transfer between the semiconductor and electrolyte, followed by the effect of illumination and charge recombination on charge transport. Finally, we discuss the effects due to surface trap states and the relation of the theoretical expressions and experimental results.

Keywords: semiconductor, electrolyte, redox, solar energy, photoelectrochemical, hydrogen

1. Introduction

The transition to renewable energy sources is recognized as one of the greatest societal challenges of the twenty-first century, and it will have a major impact on climate, environment, and economy [1]. The major bottleneck for broader utilization of renewable energy is missing large-scale and long-term energy storage technologies [2, 3]. In this respect, fossil fuels make up 85% of the worldwide energy consumption, and they have an order of magnitude higher energy density than lithium-ion batteries [4]. Hence, to replace fossil fuels and also enable a low-carbon economy in the long term, an alternative route to produce fuels exclusively from the renewable resources is sought for.

In the last two decades, a lot of research effort was given to produce renewable fuels from solar energy and/or CO_2 [5–7]. Hydrogen is one of the leading renewable fuels with important impact on a variety of industrial processes [8, 9], and European Union recognized its importance in 2020 and put an ambitious "Hydrogen Strategy for Climate-Neutral Europe" (https://ec.europa.eu/energy/sites/ener/files/hydrogen_strategy.pdf). From the variety of approaches for the renewable hydrogen production [6], great research activity is given to the photoelectrochemical (PEC) hydrogen production. For overview of the PEC water splitting principles, material requirements, characterization methods, and device architectures, we suggest the recent handbooks [10–12].

This chapter discusses theory related to the core functionality of the PEC hydrogen production or PEC CO_2 reduction, which lies in the separation of the photogenerated charges by the electric field created by the semiconductor/electrolyte interface. The physical processes at the typical n-type photoanode/electrolyte junction in the alkaline electrolyte are shown in **Figure 1a**. When a semiconductor is immersed in a solution, a charge transfer occurs at the interface because of the difference in the electrochemical potential of the two phases, creating a built-in electrostatic field, which separates electronic charges. When the light is absorbed in the semiconductor, it excites electrons and holes to their respective bands, and these are separated by the built-in electrostatic field. Holes move to the electrolyte, consuming OH^-, and oxidize water to oxygen. Electrons move away from the electrolyte and are sent through the external wire to the metal counterelectrode with an additional bias voltage, where they reduce water to hydrogen and OH^-. The exchange of OH^- between the photoanode and the counterelectrode is completed by ion migration in the electrolyte. Although with intensive research efforts, a single semiconductor material being able to do overall PEC water splitting (two half-reactions) has not yet been identified.

Therefore, a natural strategy is to distribute the two half-reactions to two semiconductor photoelectrodes (similar to Z-scheme in natural photosynthesis) [13], where wider-bandgap n-type semiconductor photoanode (1.8–2.4 eV) drives oxygen evolution reaction, and a smaller-bandgap semiconductor photocathode (1.0–1.5 eV) placed behind it drives hydrogen evolution reaction, **Figure 1b**. Such tandem configuration makes optimum usage of the full solar spectrum, leading to higher efficiencies and a better ability to cope with the fluctuating illumination conditions.

The PEC hydrogen production is currently at the lab scale and has the potential to compete if solar-to-hydrogen (STH) efficiencies over 10–15% and stability for over 10 years are considered [14, 15]. The highest reported STH efficiencies for PEC water splitting, **Figure 2**, are 14–19% for III–V tandems [16, 17], 10–14% for silicon-based multijunctions [18], and 1–3% for oxide-based devices [13].

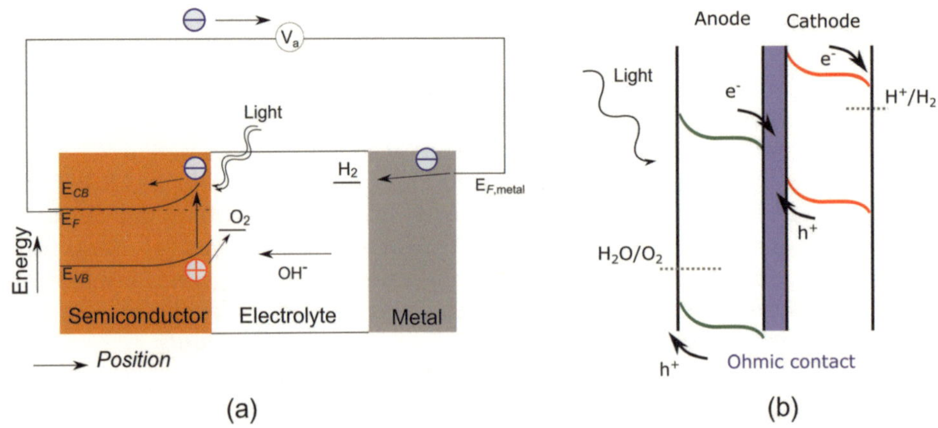

Figure 1.
(a) Schematic diagram of processes in the photoanode during three-electrode investigation for water splitting in alkaline electrolyte. (b) Sketch of the tandem PEC cell with large-bandgap photoanode and small-bandgap photocathode connected via an ohmic contact.

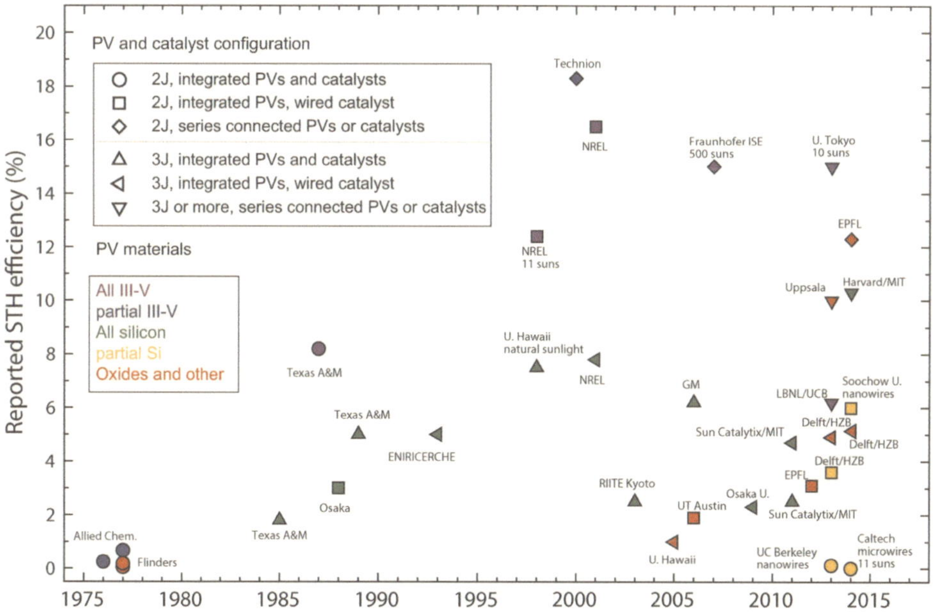

Figure 2.
Reported STH efficiencies as a function of year and sorted by the number of tandem photovoltaic junctions used (2 or 3). The fill color represents the semiconductor materials used in the photovoltaic portion of the device. All STH conversion efficiencies are as reported in the original publications. Reprinted with the permission from the Royal Society of Chemistry [13].

2. Semiconductor/liquid junction

When semiconductor is immersed in liquid electrolyte containing a redox couple, electron transfer between occupied (empty) state in the semiconductor and empty (occupied) state of the ion species in the solution is possible, respectively. This charge transfer will continue until an equilibrium is reached when Fermi level of the semiconductor E_F equals to the redox Fermi level in the electrolyte E_{redox}. Band edge positions of some semiconductors with respect to the redox potentials for water splitting are shown in **Figure 3**. Under equilibrium conditions, semiconductor region close to semiconductor/electrolyte interface (SEI) is depleted of the majority carriers, which is counterbalanced by the ions adsorbed at the semiconductor from the electrolyte side.

We start by describing the charge transfer in the semiconductor in the semiclassical approach. The resulting electrostatic potential $\phi(x)$, electron and hole concentrations $n(x), p(x)$ in the semiconductor can be calculated by solving the Poisson equation together with the carrier continuity equations

$$\frac{d^2\phi}{dx^2} = -\frac{\rho(x)}{\varepsilon_0\varepsilon_r}, \tag{1}$$

$$-\frac{1}{q}\frac{dj_e}{dx} = G(x) - R(x), \tag{2}$$

$$\frac{1}{q}\frac{dj_h}{dx} = G(x) - R(x), \tag{3}$$

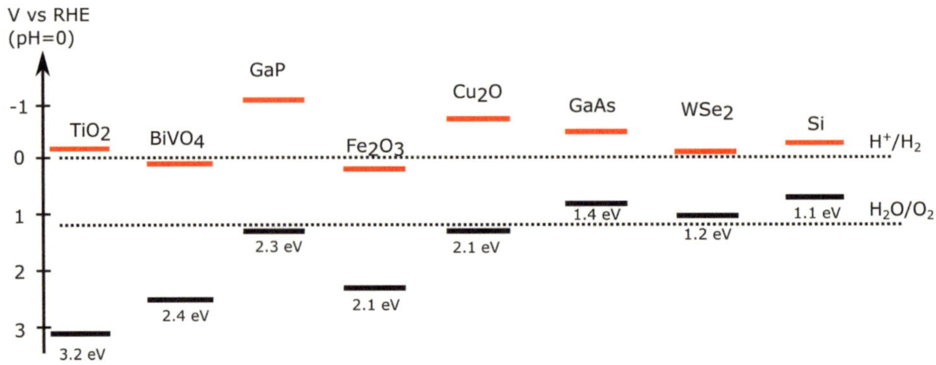

Figure 3.
Band edge positions of various semiconductors at pH = 0 (conduction band is shown in red, valence band in black). Voltage is reported with respect RHE and redox potentials of water splitting are shown as dashed lines. Bandgap energy is shown below the valence band position.

where $\rho(x) = q(N_D - n(x) + p(x))$ is the charge density, N_D is the concentration of the ionized donors for n-type semiconductor (full ionization is assumed), and $\varepsilon_0, \varepsilon_r$ denote permittivity of vacuum and relative permittivity of the semiconductor. The generation rate of the carriers is labeled $G(x)$ and the recombination rate $R(x)$. The electron and hole currents are given by the drift and diffusion terms

$$j_e = qD_e \frac{\mathrm{d}n(x)}{\mathrm{d}x} - q\mu_e n(x) \frac{\mathrm{d}\phi(x)}{\mathrm{d}x} \tag{4}$$

and

$$j_h = -qD_h \frac{\mathrm{d}p(x)}{\mathrm{d}x} - q\mu_h p(x) \frac{\mathrm{d}\phi(x)}{\mathrm{d}x}, \tag{5}$$

where electron and hole mobility are denoted μ_e, μ_h.

Usually, a space charge region (SCR) approximation is used to solve the stand-alone Poisson equation (without continuity equations), **Figure 4**, by assuming that a constant charge density exists in the SCR of width w

$$\rho(x) = qN_D, 0 < x < w \tag{6}$$

and zero charge density exists outside of the SCR. The boundary condition of the vanishing electrostatic field at the edge of SCR gives $\frac{\mathrm{d}\phi}{\mathrm{d}x}(w) = 0$, and furthermore, we choose arbitrary value $\phi(w) = 0$. In this way, the Poisson equation is integrated to give $\phi(x)$ inside SCR

$$\phi(x) = -\frac{qN_D}{2\varepsilon_0\varepsilon_r}(w - x)^2. \tag{7}$$

The width of the SCR is given by

$$w = \sqrt{\frac{2\varepsilon_0\varepsilon_r V_{sc}}{qN_D}}. \tag{8}$$

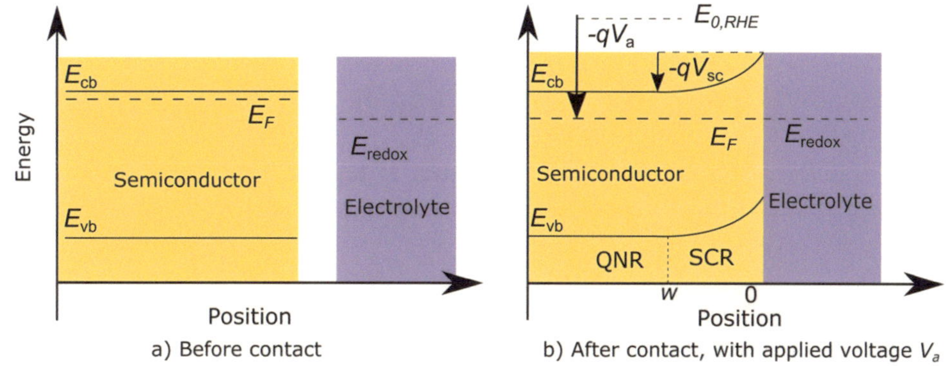

Figure 4.
(a) Energy diagram of stand-alone semiconductor and electrolyte, before contacting them. (b) Energy diagram of semiconductor and electrolyte after contact and application of voltage V_a with respect to reversible hydrogen electrode $E_{0,RHE}$.

The potential drop in the SCR region can be altered by applying the voltage V_a to the semiconductor back contact

$$V_{sc} = V_a - V_{fb}, \tag{9}$$

where for the flatband voltage $V_a = V_{fb}$, the bands of the semiconductor become flat and $V_{sc} = 0$. The semiconductor energy bands remain usually pinned at the interface to the electrolyte (with or without redox couple) as measured by the impedance spectroscopy [19], and this is also valid for $\phi(x)$, Eq. (7). Therefore, position of the semiconductor band edges can be given on the traditional electrochemical energy scale, for example, relative to the reversible hydrogen electrode (RHE) [20]. Due to strong interaction of semiconductors with water in aqueous elecrolytes, applying the voltage bias to the semiconductor leads only to the change of the electrostatic potential across the space charge layer V_{sc} (the potential across the Helmholtz double layer remains constant). The energies of the conduction band and valence band E_{cb}, E_{vb} are given by

$$E_{cb}(x) = -\chi - q\phi(x), E_{vb}(x) = E_{cb}(x) - E_g, \tag{10}$$

and the quasi-Fermi level of electrons and holes E_{Fn}, E_{Fp}

$$E_{Fn} = E_{cb} + kT\ln\left(\frac{n}{N_c}\right), E_{Fp} = E_{vb} - kT\ln\left(\frac{p}{N_v}\right), \tag{11}$$

where the electron affinity is labeled χ, symbol k denotes the Boltzmann constant, T is the absolute temperature. We remark that in our discussion, no electronic states at the surface of the semiconductor are assumed for simplicity. The electron and hole concentrations at the SEI, n_s and p_s, are given by the Boltzmann distribution

$$n_s = n_0 \exp\left(-\frac{qV_{sc}}{kT}\right) \tag{12}$$

and

$$p_s = p_0 \exp\left(+\frac{qV_{sc}}{kT}\right). \tag{13}$$

The equilibrium electron (hole) concentration is denoted as n_0 (p_0). The equilibrium hole concentration p_0 is given by the expression

$$p_0 = \frac{N_c N_v \exp\left(-\frac{E_g}{kT}\right)}{N_D}, \tag{14}$$

where the effective densities of states in the CB and VB are denoted N_c and N_v and the bandgap energy is E_g.

2.1 In the dark

Turning now our attention to the redox couple, the energy values of *Red* (*Ox*) species in the solution E_{red}, E_{ox} can be described with the solvation state model introduced by Gerischer [21]. Redox couple is a pair of two ions (*Red* is the reduced species, *Ox* is the oxidized species), which can interchange electrons

$$Red - e^- - > Ox \tag{15}$$

The Nernst equation describes the electrochemical potential of electrons in the redox couple, which is equivalent to the Fermi level of the redox couple E_{redox} when the same reference is used for the semiconductor and redox system [22].

$$E_{redox} = E^0_{redox} + kT \ln\left(\frac{c_{ox}}{c_{red}}\right), \tag{16}$$

where the reference redox level is denoted E^0_{redox}, and c_{red}, c_{ox} are the concentrations of the $Red(Ox)$ species. Herein, we assume weak interaction of the semiconductor with the redox couple and equal concentrations of *Red* and *Ox* species. The addition (removal) of an electron to (from) the *Ox* (*Red*) species is accompanied by the increase (decrease) of the energy equal to the reorganization energy λ (typically around 1 eV). Taking into account the rotation and motion of the solvent ions, probability distributions W_{ox}, W_{red} of the energy states around the mean values E_{ox}, E_{red} are obtained and shown in **Figure 5**

$$W_{ox}(E) = W^0 \exp\left(-\frac{(E - E_{redox} + \lambda)^2}{4kT\lambda}\right), \tag{17}$$

$$W_{red}(E) = W^0 \exp\left(-\frac{(E - E_{redox} - \lambda)^2}{4kT\lambda}\right), \tag{18}$$

where $W^0 = (4kT\lambda)^{-1/2}$ is a constant to provide unit integrated probability over the whole energy spectrum. If E_{redox} is closer to the conduction (valence) band edge of the semiconductor, electron exchange between the conduction (valence) band and

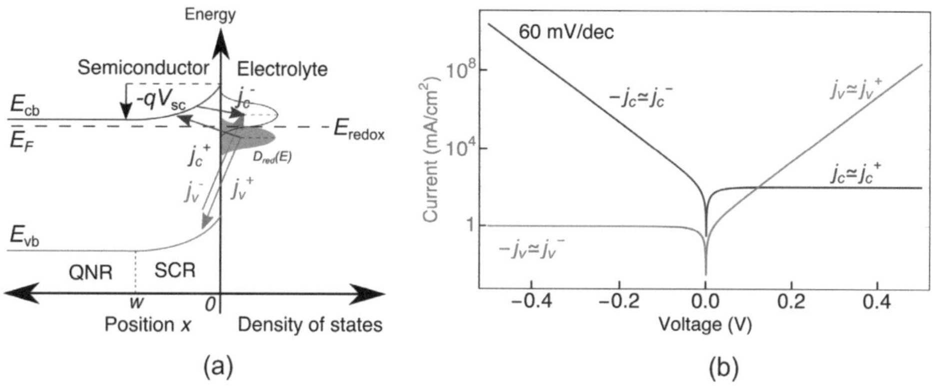

Figure 5.
N-type semiconductor in contact with a redox couple in the electrolyte with E_{redox} close to E_F. (a) Energy diagram along with distribution of the occupied states in the electrolyte (painted in green color). The conduction (valence) band energy is denoted E_{cb}, E_{vb}, and the Fermi energy is labeled E_F. (b) The current-voltage curves in the logarithmic scale for parameters $j_v^\circ = 1\ mA/cm^2$ and $j_c^\circ = 100\ mA/cm^2$. The negative currents were shown with a positive value in the logarithmic plot.

redox couple is energetically preferred, which is a consequence of small loss of electron energy when both energy states have similar energy. The energy distribution functions are proportional to the concentration of the respective species and probability distribution

$$D_{ox} = c_{ox} W_{ox}, \tag{19}$$

$$D_{red} = c_{red} W_{red}. \tag{20}$$

The rate of electron transfer from the semiconductor to the electrolyte is given by the integral of the probability of the transition between these quantum states over all electron energies

$$\int_E f(E)\rho(E)D_{ox}(E)\mathrm{d}E, \tag{21}$$

where $f(E)$ denotes the Fermi function in the semiconductor, $\rho(E)$ denotes the energy distribution in the semiconductor, and $W_{ox}(E)$ is the distribution of empty states in the electrolyte. For the electron transfer from the conduction band to the electrolyte (cathodic current, superscript-), the latter expression becomes

$$j_c^- = qc_{ox}\int_{E_c}^{\infty} n_s \exp\left(-\frac{(E - E_{redox} + \lambda)^2}{4kT\lambda}\right), \tag{22}$$

with the electron concentration at the semiconductor-electrolyte interface n_s. Typically, by assuming a small overlap of the energy states in the semiconductor and electrolyte (within $1kT$ of the conduction band), the latter integral is simplified to

$$j_c^- = qk_c^- n_s c_{ox}, \tag{23}$$

with k_c^- is the rate constant containing the exponential term (potential independent), and q is the electronic charge. The current j_c^+ corresponding to the opposite process of the electron transfer from the electrolyte to the semiconductor (anodic current, superscript +) is calculated similarly

$$j_c^+ = qk_c^+ N_c c_{red}, \tag{24}$$

where we have used that the density of empty states in semiconductor equals N_C, and the distribution of the occupied states is now proportional to c_{red}.

In a similar way, the cathodic current from the valence band is potential-independent

$$j_v^- = qk_v^- N_v c_{ox} \tag{25}$$

whereas the anodic current from the valence band depends on the potential through p_s

$$j_v^+ = qk_v^+ p_s c_{red}. \tag{26}$$

Further simplification of the expressions for currents is derived by considering the balanced equilibrium charge transfer with the equilibrium conduction band current j_c^0

$$j_c^+ = j_c^- = j_c^0, \tag{27}$$

and equilibrium valence band current j_v^0

$$j_v^+ = j_v^- = j_v^0. \tag{28}$$

In equilibrium $n_s = n_s^0, p_s = p_s^0$. Henceforth, the net current through the conduction band from the semiconductor to the electrolyte is obtained by subtracting the cathodic current from the anodic current

$$j_c = j_c^+ - j_c^- = -j_c^0\left(\frac{n_s}{n_s^0} - 1\right) = -j_c^0\left[\exp\left(-\frac{qV_{sc}}{kT}\right) - 1\right] \tag{29}$$

and accordingly

$$j_v = j_v^+ - j_v^- = j_v^0\left(\frac{p_s}{p_s^0} - 1\right) = j_v^0\left[\exp\left(\frac{qV_{sc}}{kT}\right) - 1\right] \tag{30}$$

Our notation is compatible with the traditional electrochemical notation of the positive anodic current and the negative cathodic current, **Figure 5c**. We remark, however, that the conduction or valence band current is positive (negative) for the cathodic (anodic) bias voltage.

To illustrate the profile of the current-voltage curves, we have taken a small j_v^0 and large j_c^0, which is expected for E_{redox} positioned close to the conduction band of n-type semiconductor. For a positive voltage (reverse bias voltage), the anodic current j_c^+ of electrons from electrolyte to the conduction band is approximately constant. For the increasing negative voltage (forward bias voltage), more electrons (majority carriers) become available, and hence, the cathodic current j_c^- of electrons from the conduction band to the electrolyte rises exponentially (rectification). The valence band currents are much smaller in magnitude and follow an exponential rise of j_v^+ for the reverse bias and a constant value of j_v^- for the forward bias (opposite trend as compared with the conduction band current). Generally, a redox reaction at the semiconductor-eletrolyte interface can either receive electrons from the semiconductor or send electrons to the semiconductor. Electrons are majority (minority) carriers in the n-type (p-type) semiconductor, respectively, and hence, the situation with the relative magnitude of the currents across the interface becomes interesting when minority carriers are created by light excitation and can completely alter the current-voltage behavior. The total current at the semiconductor electrode with the Tafel slope 60 mV/dec

$$j = j_c + j_v = -j_c^0\left[\exp\left(-\frac{qV_{sc}}{kT}\right) - 1\right] + j_v^0\left[\exp\left(\frac{qV_{sc}}{kT}\right) - 1\right]. \tag{31}$$

partially resembles the Butler-Volmer equation for the current at the metal electrode with Tafel slope 120 mV/dec

$$j = j_0\left[\exp\left(\frac{qV}{2kT}\right) - \exp\left(-\frac{qV}{2kT}\right)\right]. \tag{32}$$

The difference between the metal and semiconductor current is caused by the potential drop falling over the Helmholtz layer in the metal, while the potential drops over SCR in the semiconductor. Furthermore, the exchange current density is much smaller for the semiconductor than for a metal since metals have much higher density of states near Fermi level than semiconductors.

For the n-type semiconductor and electrons (majority carriers) being transferred from the electrolyte to the semiconductor, it is usually described in the literature that holes (minority carriers) are transported from the semiconductor to the electrolyte and minority carrier reaction proceeds at the interface. In this situation, the anodic current is dominated by j_v^+ since j_c^- is essentially constant (due to a constant number of the empty states in the conduction band N_C).

If electrons (majority carriers for n-type) are transferred from the n-type semiconductor to the electrolyte, it is usually termed majority carrier reaction, and hence, the cathodic current is dominated by j_c^- since j_v^- is essentially constant (due to a constant number of occupied states in the valence band N_V).

2.2 Upon illumination

In the preceding section, we presented anodic and cathodic current exchange with the electrolyte in the dark, and we neglected the carrier recombination. Magnitude of the first-order recombination is governed by the population of very few available minority carriers. Additionally, population of the minorites cannot be altered by the voltage bias, as only majority carriers can be injected by the voltage bias [23].

When a light of sufficient energy strikes the semiconductor, the generated electron-hole pairs contribute to the electron exchange with the redox couple or recombine. Due to carrier recombination, the currents calculated in the preceding section cannot be used without further considerations. Here we sketch the development due to Reichman [24], which is instructive but simple enough for tractability. The total current under illumination is composed of the conduction band and valence band currents

$$j = j_c + j_v. \tag{33}$$

The conduction band current, Eq. (30), remains the same under illumination as the electron population is not substantially altered upon illumination. The valence band current, Eq. (30), holds when the hole concentration at the surface p_s is calculated under illumination

$$j_v = j_v^0 \left(\frac{p_s}{p_s^0} - 1 \right). \tag{34}$$

To obtain p_s, it is assumed that quasi-equilibrium of holes is valid even under illumination

$$p(x) = p_w \exp\left(+ \frac{qV_{sc}}{kT} \right) \tag{35}$$

and the stand-alone hole continuity equation, Eq. (3), is solved in the neutral region of the semiconductor, where electric field can be neglected. This simplification leads to the diffusion equation valid when hole diffusion length L_h is much smaller than the neutral region thickness

$$D_h \frac{d^2 p}{dx^2} - \frac{p - p_0}{\tau_h} + I_0 \alpha \exp(-\alpha x) = 0. \tag{36}$$

The hole diffusion constant is written as D_h, hole recombination lifetime τ_h (first-order recombination), absorption coefficient α, and incoming monochromatic photon flux from the electrolyte side I_0 (assuming Lambert-Beer conditions). The analytic solution with the boundary conditions $p(\infty) = p_0$ and $p(w) = p_w$ is

$$p = p_0 + \frac{\alpha I_0 \tau_h \exp(-\alpha x)}{1 - \alpha^2 L_h^2} + K \exp(-x/L_h), \tag{37}$$

where $L_h^2 = D_h \tau_h$ and the constant K

$$K = \exp(w/L_h) \left(p_w - p_0 - \frac{\alpha I_0 \tau_h \exp(-\alpha w)}{1 - \alpha^2 L_h^2} \right). \tag{38}$$

The diffusion current at the edge of the neutral region $x = w$ is

$$j_{diff} = -j_0 \left(\frac{p_w}{p_0} - 1 \right) + \frac{q I_0 \alpha L_h \exp(-\alpha w)}{1 + \alpha L_h}, \tag{39}$$

with the saturation current density denoted $j_0 = \frac{q p_0 D_h}{L_h}$.

In the space-charge region, we simply assume all photogenerated charges are separated by the electric field of the SEI, and there is no recombination, hence, the hole current to the electrolyte is

$$j_{sc} = q I_0 [1 - \exp(-\alpha w)]. \tag{40}$$

The valence band current is then sum of the currents in the neutral and the space-charge region for both anodic and cathodic bias voltage

$$j_v = j_{diff} + j_{sc} = \frac{j_g - j_0 \exp\left(-\frac{qV_{sc}}{kT}\right)}{1 + \frac{j_0}{j_v^0} \exp\left(-\frac{qV_{sc}}{kT}\right)}. \tag{41}$$

The term

$$j_g = j_0 + qI_0 \left[1 - \frac{\exp(-\alpha w)}{1 + \alpha L_h}\right] \tag{42}$$

represents the generation current. We remark that the Gartner photocurrent j_G corresponds to the second term in the last equation

$$j_G = qI_0 \left[1 - \frac{\exp(-\alpha w)}{1 + \alpha L_h}\right] \tag{43}$$

and was derived by assuming the ideal collection of holes within the SCR, no recombination in SCR and $p_w = 0$ [25]. The photocurrent onset voltage for Gartner equation overlaps with the flatband voltage V_{fb} of the SEI, **Figure 6b**. There are two current constants j_0, j_v^0 in Reichman photocurrent Eq. (41), which govern the profile of JV curve.

Without illumination ($I_0 = 0$), the Reichman dark current includes recombination in neutral region and equals

$$j_v^{dark} = \frac{j_0 - j_0 \exp\left(-\frac{qV_{sc}}{kT}\right)}{1 + \frac{j_0}{j_v^0} \exp\left(-\frac{qV_{sc}}{kT}\right)}. \tag{44}$$

Photocurrent j_v has onset delayed more anodic with respect to Gartner photocurrent j_G for decreasing (slower) rate of hole transfer to electrolyte j_v^0, see linear-scale current-voltage plot, **Figure 6a**. This is expected behavior as holes queue at SEI due to their slow transfer to electrolyte. For large anodic bias, Reichman photocurrent j_v approaches Gartner photocurrent j_G as recombination becomes the limiting process. Under large anodic bias voltage ($V_{sc} > 0$), the dark current approaches value j_0 and the photocurrent reaches j_G, **Figure 6**. Under large cathodic bias voltage ($V_{sc} < 0$), both dark current and photocurrent approach j_v^0.

For intermediate anodic bias voltage, both the dark current and photocurrent scale as $\sim \exp\left(-\frac{qV_{sc}}{kT}\right)$ (Tafel slope 60 mV/dec) and minority carrier recombination and transfer govern the kinetics similar to the pn-junction or semiconductor/metal contact. The range of the intermediate anodic bias ($0 < V_{sc} < V_c$) with approximate slope 60 mV/dec increases with the decreasing value of j_v^0 (rate of hole current transfer at the SEI). The critical voltage bias V_c can be derived from the condition that the terms in the denominator become equal and is given by [26, 27].

$$V_c = \frac{kT}{q} \ln \frac{j_0}{j_v^0}. \tag{45}$$

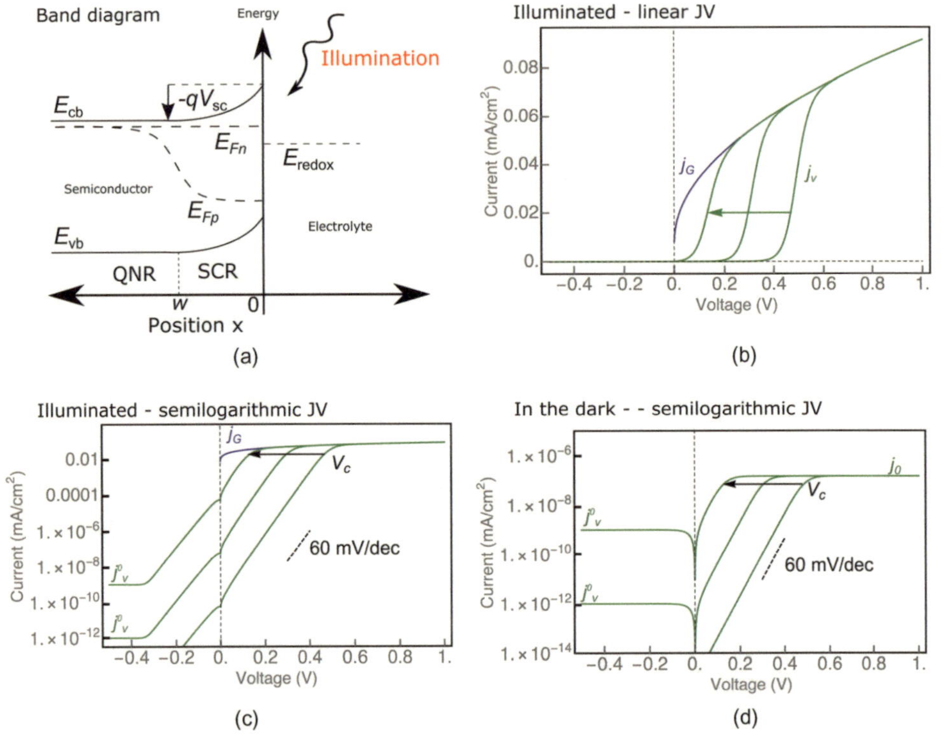

Figure 6.
(a) Schematic of the illuminated SEI. (b)–(d) Current-voltage Eq. (41) for baseline parameters in ***Table 1*** *and several values of $j_v^0 = 10^{-15}, 10^{-12}, 10^{-9}$ mA/cm² (direction of arrow, placed at the corresponding critical voltage V_c) in (b) linear plot and (c) semi-logarithmic plot and (d) in the dark (without illumination). For the cathodic (negative) bias voltage, sign of the current is reversed to enable plot in the semilogarithmic scale.*

For baseline parameters and $j_v^0 = 10^{-15}, 10^{-12}, 10^{-9}$ mA/cm², we get $V_c = 0.49, 0.31, 0.13$ V, and these correspond well with plots in **Figure 6**.

For completeness, we remark that SCR recombination can be easily incorporated in the treatment of Reichman, whereas we decided not to include this term in our plots [24]. The recombination current in the SCR is

$$j_{sc}^R = Q\sqrt{\frac{p_s}{p_{s0}}}, \tag{46}$$

where these expressions are used

$$Q = \frac{\pi k T n_i w \exp\left(-\frac{qV_{sc}}{2kT}\right)}{4\tau_h V_{sc}}, \tag{47}$$

$$\sqrt{\frac{p_s}{p_{s0}}} = \frac{-Q + \sqrt{Q^2 + 4UW}}{2U}, \tag{48}$$

$$U = j_v^0 + j_0 \exp\left(-\frac{qV_{sc}}{kT}\right), \tag{49}$$

$$W = j_v^0 + j_G. \tag{50}$$

Parameter	Value	Description
E_g (eV)	1.2	Bandgap energy
ε_r	10	Relative permittivity
N_D (cm^{-3})	10^{17}	Donor concentration
N_c (cm^{-3})	10^{20}	Effective density of states in CB
N_v (cm^{-3})	10^{20}	Effective density of states in CB
L (nm)	100	Semiconductor thickness
α (cm^{-1})	10^3	Absorption coefficient of the semiconductor
L_h (nm)	10	Hole diffusion length
τ_h (ps)	1	Hole recombination lifetime
I_0 (cm^{-2}/s)	$5{\cdot}10^{16}$	Incoming monochromatic photon flux
V_{fb} (V vs. RHE)	0	Flatband voltage

Table 1.
Baseline material properties of the N-type semiconductor photoelectrode used in this chapter.

For very large j_v^0, the cathodic current can be diffusion-controlled, and we do not discuss this point here. Around the flatband voltage, the conduction band current, Eq. (30), might dominate the total current due to dark conduction band current, **Figure 5b**. The presence of bulk trap states in the semiconductor [23] or catalyst layer on the semiconductor [28, 29] was recently treated in model systems, while these are not considered herein.

The relation of the theoretical and measured photocurrent-voltage curves is discussed next. The conditions assumed for the Gartner photocurrent (fast electron transfer to the electrolyte, negligible recombination in the SCR or at the surface) are valid for high-quality semiconductors such as silicon, GaAs, GaP, **Figure 7a**, and the measured photocurrent onset voltage for GaP is $\sim$0.1 V larger than the flatband voltage (which marks the onset voltage of the Gartner photocurrent). This is not the case for most metal oxide photoelectrodes with short diffusion length of minority carriers and slow electron transfer to electrolyte. Taking iron oxide Fe_2O_3 as an example, measured photocurrent onset voltage is $\sim$0.4 V or more anodic of the flatband voltage, and the photocurrent rises slowly in the beginning (so called S-shape) in contrast to the steep rise predicted by the Gartner equation, **Figure 7b**. Some efforts lead to reproducing of the photocurrent S-shape by considering the surface recombination [32], but in general, spectroscopic methods are needed to provide an additional evidence for the correctness of the model and the extracted parameters from the photocurrent response.

The complications that arise due to recombination in the surface states were treated by several studies [32–35], and here we briefly outline theory by Peter [36, 37]. For the surface trap at the SEI with concentration N_s located E_s above the Fermi level of the bulk semiconductor, the trap occupancy in the dark is

$$f_0 = \frac{1}{1 + \exp\left(\frac{E_s - qV_{sc}}{kT}\right)} \tag{51}$$

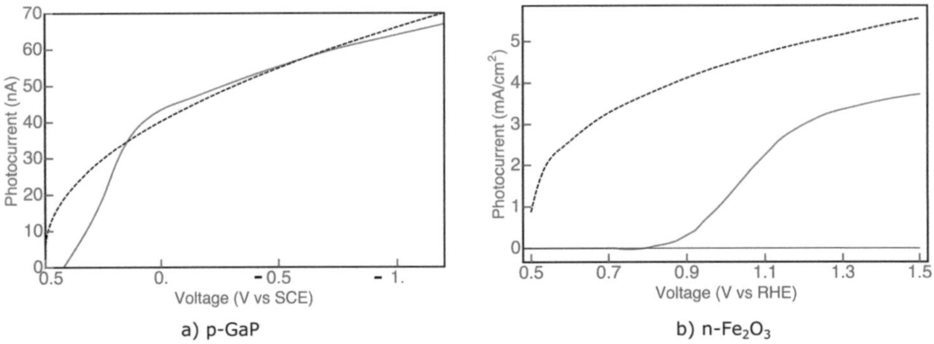

Figure 7.
(a) Measured photocurrent-voltage curves (green color) for p-GaP in 0.5 mol H_2SO_4 recorded for the low-level monochromatic illumination at fixed wavelengths, the data were taken from [30]. We remark that voltage is measured against the saturated calomel electrode (SCE), and negative voltage is to the right. The Gartner photocurrent is shown in dashed black line. (b) Photocurrent-voltage curve for n-Fe_2O_3 (hematite) in NaOH solution and simulated AM1.5G illumination, the data were taken from [31]. The corresponding Gartner photocurrent is shown in dashed black line.

and the trap occupancy under illumination is

$$f = \frac{-B - \sqrt{B^2 - 4AC}}{2A}, \tag{52}$$

where constants A, B, C are

$$A = k_n k_p n_s N_s^2, B = -k_n N_s j_G - A(1 + f_0) - k_0 k_p n_s N_s \tag{53}$$

and

$$C = k_n N_s j_G + A f_0 + k_0 k_p n_s f_0. \tag{54}$$

Henceforth, the photocurrent voltage is described by the relation

$$j_s = j_G - q k_n n_s N_s (f - f_0). \tag{55}$$

For the increasing concentration of the surface states, the photocurrent rises steeply as the surface state occupation reaches unity when the Fermi level of the semiconductor moves through the surface state energy level, **Figure 8a**.

The charging of the surface states influences the distribution of electrostatic potential in the semiconductor. The potential drop in the Helmholtz layer can be written as

$$\delta V_H = \frac{q N_s (1 - f_0)}{C_H}, \tag{56}$$

when the vacant surface state is assumed to carry a positive charge, and C_H is the capacitance of the Helmholtz layer. The increasing surface state concentration leads to more band-edge unpinning shown as flattening in **Figure 8b**.

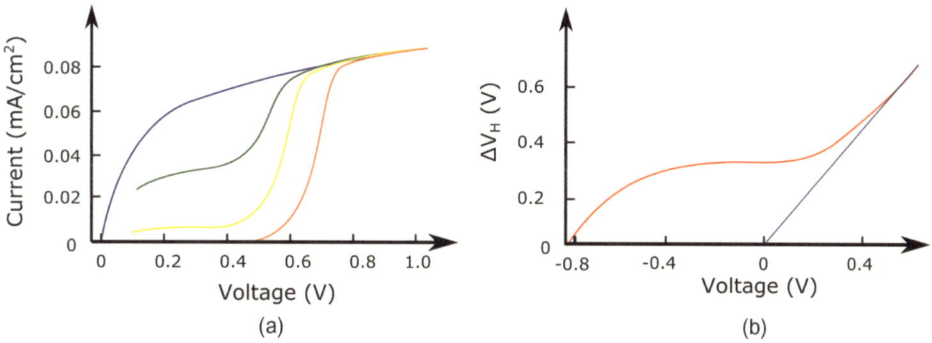

Figure 8.
(a) Photocurrent-voltage curves for surface recombination via surface state located E_s =0.3 eV above the bulk Fermi level and $N_s = 10^{11}, 10^{12}, 10^{14}\ cm^{-2}$. (b) Potential drop in the Helmholtz layer for the same N_s as in (a).

3. Conclusions

This chapter provided introduction to the topic of charge transport in the dark and illuminated semiconductor/liquid junction in order to understand behavior of the ideal SEI. The level to which real semiconductor electrodes mimic such ideal SEI can be very broad. First, we introduced basic considerations of the semiclassical charge transport equations in the semiconductor. Second, the interaction of the semiconductor with redox system in the electrolyte was described, along with the energetics of the electron transfer between semiconductor and redox system. Third, the importance of recombination kinetics in the semiconductor was considered upon illumination, and classical photocurrent formulas due to Gartner and Reichman were discussed. Finally, the effect of surface state occupancy on the photocurrent-voltage curves was described.

Acknowledgements

Funding for this work was partially covered by grant of APVV-20-0528.

Conflict of interest

The authors declare no conflict of interest.

Author details

Peter Cendula
Institute of Aurel Stodola, Faculty of Electrical Engineering and Information Technology, University of Zilina, Komenskeho, Mikulas, Slovakia

*Address all correspondence to: peter.cendula@feit.uniza.sk

References

[1] A Roadmap for Moving to a Competitive Low Carbon Economy in 2050. Technical Report. 2011. Available from: http://ec.europa.eu/clima/policies/roadmap/documentation_en.htm

[2] Lewis NS, Nocera DG. Powering the planet: Chemical challenges in solar energy utilization. Proceedings of the National Academy of Sciences. 2006; **103**(43):15729-15735. DOI: 10.1073/pnas.0603395103

[3] Luo X, Wang J, Dooner M, Clarke J. Overview of current development in electrical energy storage technologies and the application potential in power system operation. Applied Energy. 2015; **137**:511-536. DOI: 10.1016/j.apenergy.2014.09.081

[4] Züttel A, Remhof A, Borgschulte A, Friedrichs O. Hydrogen: The future energy carrier. Philosophical Transactions of the Royal Society A: Mathematical, Physical and Engineering Sciences. 1923;**2010**(368):3329-3342. DOI: 10.1098/rsta.2010.0113

[5] Kumar B, Llorente M, Froehlich J, Dang T, Sathrum A, Kubiak CP. Photochemical and photoelectrochemical reduction of CO_2. Annual Review of Physical Chemistry. 2012;**63**(1):541-569. DOI: 10.1146/annurev-physchem-032511-143759

[6] Turner J, Sverdrup G, Mann MK, Maness P-C, Kroposki B, Ghirardi M, et al. Renewable hydrogen production. International Journal of Energy Research. 2008;**32**(5):379-407. DOI: 10.1002/er.1372

[7] White JL, Baruch MF, Pander JE, Hu Y, Fortmeyer IC, Park JE, et al. Light-driven heterogeneous reduction of carbon dioxide: Photocatalysts and photoelectrodes. Chemical Reviews. 2015;**115**(23):12888-12935. DOI: 10.1021/acs.chemrev.5b00370

[8] Hart D, Lehner F, Rose R, Lewis J, Klippenstein M. Fuel Cell Industry Review 2017. 2017. Available from: https://fuelcellindustryreview.com/

[9] Staffell I, Scamman D, Velazquez Abad A, Balcombe P, Dodds PE, Ekins P, et al. The role of hydrogen and fuel cells in the global energy system. Energy & Environmental Science. 2019;**12**(2):463-491. DOI: 10.1039/C8EE01157E

[10] Giménez S, Bisquert J. Photoelectrochemical Solar Fuel Production: From Basic Principles to Advanced Devices. Berlin: Springer; 2016

[11] Van De Krol R, Grätzel M. Photoelectrochemical Hydrogen Production. New York, USA: Springer; 2011

[12] Lewerenz H-J, Peter L. Photoelectrochemical Water Splitting: Materials, Processes and Architectures. Cambridge, UK: Royal Society of Chemistry; 2013

[13] Ager JW, Shaner M, Walczak K, Sharp ID, Ardo S. Experimental demonstrations of spontaneous, solar-driven photoelectrochemical water splitting. Energy & Environmental Science. 2015;**8**:2811-2824. DOI: 10.1039/C5EE00457H

[14] James BD, Baum GN, Perez J, Baum KN, One Virginia Square. Technoeconomic analysis of photoelectrochemical (PEC) hydrogen production. Directed Technologies Inc., Department of Energy Contract

GS-10F-009J Technical Report. 2009. Available from: http://www1.eere.energy.gov/hydrogenandfuelcells/pdfs/pec_technoeconomic_analysis.pdf

[15] Pinaud BA, Benck JD, Seitz LC, Forman AJ, Chen Z, Deutsch TG, et al. Technical and economic feasibility of centralized facilities for solar hydrogen production via photocatalysis and photoelectrochemistry. Energy & Environmental Science. 2013;**6**(7): 1983-2002. DOI: 10.1039/C3EE40831K

[16] May MM, Lackner D, Ohlmann J, Dimroth F, van de Krol R, Hannappel T, et al. On the benchmarking of multi-junction photoelectrochemical fuel generating devices. Sustainable Energy & Fuels. 2017;**1**(3):492-503. DOI: 10.1039/C6SE00083E

[17] Verlage E, Hu S, Liu R, Jones RJR, Sun K, Xiang C, et al. A monolithically integrated, intrinsically safe, 10% efficient, solar-driven water-splitting system based on active, stable earth-abundant electrocatalysts in conjunction with tandem III–V light absorbers protected by amorphous TiO_2 films. Energy & Environmental Science. 2015; **8**:3166-3172. DOI: 10.1039/C5EE01786F

[18] Urbain F, Smirnov V, Becker J-P, Lambertz A, Yang F, Ziegler J, et al. Multijunction Si photocathodes with tunable photovoltages from 2.0 V to 2.8 V for light induced water splitting. Energy & Environmental Science. 2016; **9**:145-154. DOI: 10.1039/C5EE02393A

[19] Bisquert J, Giménez S, Bertoluzzi L, Herraiz-Cardona I. Analysis of photoelectrochemical systems by impedance spectroscopy. In: Giménez S, Bisquert J, editors. Photoelectrochemical Solar Fuel Production. Switzerland: Springer International Publishing; 2016. pp. 281-321. DOI: 10.1007/978-3-319-29641-8_6

[20] Bisquert J, Cendula P, Bertoluzzi L, Gimenez S. Energy diagram of semiconductor/electrolyte junctions. The Journal of Physical Chemistry Letters. 2013;**5**(1):205-207. DOI: 10.1021/jz402703d

[21] Gerischer H. Solar photoelectrolysis with semiconductor electrodes. In: Seraphin B, editor. Topics in Applied Physics Volume 31: Solar Energy Conversion. Berlin/Heidelberg: Springer; 1979. pp. 115-172

[22] Memming R. Semiconductor Electrochemistry. Weinheim, Germany: John Wiley & Sons; 2008

[23] Bertoluzzi L, Bisquert J. Equivalent circuit of electrons and holes in thin semiconductor films for photoelectrochemical water splitting applications. The Journal of Physical Chemistry Letters. 2012;**3**(17):2517-2522. DOI: 10.1021/jz3010909

[24] Reichman J. The current-voltage characteristics of semiconductor-electrolyte junction photovoltaic cells. Applied Physics Letters. 1980;**36**(7): 574-577. DOI: 10.1063/1.91551

[25] Gärtner WW. Depletion-layer photoeffects in semiconductors. Physical Review. 1959;**116**(1):84-87. DOI: 10.1103/PhysRev.116.84

[26] Cendula P, Sahoo PP, Cibira G, Simon P. Analytical model for photocurrent-voltage and impedance response of illuminated semiconductor/electrolyte interface under small voltage bias. The Journal of Physical Chemistry C. 2020;**124**(2):1269-1276. DOI: 10.1021/acs.jpcc.9b07244

[27] Cendula P, Sancheti A, Simon P. Model-based investigation of trap-assisted recombination in photoelectrodes for water splitting.

Advanced Theory and Simulations. 2021; **4**(12):2100312. DOI: 10.1002/adts.202100312

[28] Mills TJ, Lin F, Boettcher SW. Theory and simulations of electrocatalyst-coated semiconductor electrodes for solar water splitting. Physical Review Letters. 2014; **112**(14):148304. DOI: 10.1103/PhysRevLett.112.148304

[29] Nellist MR, Laskowski FAL, Lin F, Mills TJ, Boettcher SW. Semiconductor–electrocatalyst interfaces: Theory, experiment, and applications in photoelectrochemical water splitting. Accounts of Chemical Research. 2016; **49**(4):733-740. DOI: 10.1021/acs.accounts.6b00001

[30] Li J, Peat R, Peter LM. Surface recombination at semiconductor electrodes: Part II. Photoinduced "near-surface" recombination centres in p-GaP. Journal of Electroanalytical Chemistry and Interfacial Electrochemistry. 1984;**165**(1–2):41-59. DOI: 10.1016/S0022-0728(84)80085-6

[31] David Tilley S, Cornuz M, Sivula K, Grätzel M. Light-induced water splitting with hematite: Improved nanostructure and iridium oxide catalysis. Angewandte Chemie. 2010;**122**(36):6549-6552. DOI: 10.1002/ange.201003110

[32] Wilson RH. A model for the current-voltage curve of photoexcited semiconductor electrodes. Journal of Applied Physics. 1977;**48**(10):4292-4297. DOI: 10.1063/1.323417

[33] El Guibaly F, Colbow K. Theory of photocurrent in semiconductor-electrolyte junction solar cells. Journal of Applied Physics. 1982;**53**(3):1737-1740. DOI: 10.1063/1.331642

[34] El Guibaly F, Colbow K, Funt BL. Voltage dependence of the dark and photocurrents in semiconductor-electrolyte junctions. Journal of Applied Physics. 1981;**52**(5):3480-3483. DOI: 10.1063/1.329124

[35] Jarrett HS. Photocurrent conversion efficiency in a Schottky barrier. Journal of Applied Physics. 1981;**52**(7):4681-4689. DOI: 10.1063/1.329351

[36] Peter LM, Li J, Peat R. Surface recombination at semiconductor electrodes: Part I. Transient and steady-state photocurrents. Journal of Electroanalytical Chemistry and Interfacial Electrochemistry. 1984;**165**(1–2):29-40. DOI: 10.1016/S0022-0728(84)80084-4

[37] Peter LM, Ponomarev EA, Fermín DJ. Intensity-modulated photocurrent spectroscopy: Reconciliation of phenomenological analysis with multistep electron transfer mechanisms. Journal of Electroanalytical Chemistry. 1997;**427**(1–2):79-96. DOI: 10.1016/S0022-0728(96)05033-4

Section 2

Semiconductor Materials

Chapter 5

Temperature Dependence of Electrical Resistivity of $(III, Mn)V$ Diluted Magnetic Semiconductors

Edosa Tasisa Jira

Abstract

In this work, a theory of temperature dependence of electrical resistivity is developed, with a particular emphasis on dilute magnetic semiconductors (DMSs). The approach is based on the equation of motion of the Ruderman-Kittel-Kasuya-Yosida (RKKY) exchange interaction and considers both spin and charge disorder. The formalism is applied to the specific case of $Ga_{1-x}Mn_xAs$.Using the RKKY exchange interaction, the relaxation time τ and the exchange interaction J are calculated. Then using spin-dependent relaxation time, electrical resistivity of the material is calculated. The electrical resistivity of Mn-doped III—V DMS is decreased with increasing temperature and magnetic impurity concentration.

Keywords: RKKY interactions, electrical resistivity, spin-dependent relaxation time

1. Introduction

The motivation for this study has been to better understand semiconductors, which could be the building block for the next generation of computers. Semiconductors have been the subject of very extensive research over recent decades, because of ever more numerous and powerful applications. As conventional electronics, semiconductors are more appropriate than normal metals, since metal devices do not amplify current. Because of the complementary properties of semiconductor and ferromagnetic material systems, a growing effort is directed toward studies of semiconductor-magnetic nanostructures [1, 2].

Semiconductors and magnetic materials both play essential roles in modern electronic industry. Since the applications of semiconductors and magnetic have evolved independently, it appears logical to combine their properties for possible spin-electronic applications with increased functionalities [3]. This is known as spintronics, which utilizes both charge and spin of the electrons to process and store data. This is to control the spin degree of freedom of electrons in semiconductors. The direct method to introduce spin degrees of freedom in semiconductors is to introduce magnetic ions into semiconductors. Such semiconductors are referred to as diluted magnetic semiconductors (DMSs) [4]. The curie temperature of the DMS material and whether the ferromagnetism in the DMS material originates from free carrier mediation or purely from localized magnetic dopants limit DMS practical applications for devices [5].

IntechOpen

For practical applications, increasing T_C beyond room temperature is a necessity. Thus, exploring various roots of raising the value of T_C is a great fundamental and practical significance. Carrier-mediated ferromagnetism in p-type II-VI DMS heterostructures occurs only at very low temperatures T_C typically below $2.0K$. Therefore, very significant strides have been made in developing ferromagnetic-based $III - V$ semiconductors containing Mn, which remaining ferromagnetic to much higher T_C [3, 6]. A discovery of ferromagnetic ordering in III–V DMSs with critical temperatures of annealed $Ga_{1-x}Mn_xAs$ could exceed 110 k; reaching ~191 K has renewed and greatly intensified an interest in those materials, but still too low for actual applications. This was, at least partially, related to expectations that their Curie temperatures can be relatively easily brought to room temperature range through a clearly delineated path [7–9].

For the ferromagnetic transition to occur in $Ga_{1-x}Mn_xAs$, the concentration of Mn impurities should be relatively high, x = 0.01–0.05. Such high concentration of impurities makes the study of the DMS more challenging. Moreover, the issue of solubility of the magnetic ions into the regular semiconductor was a major challenge to fabricate high-quality DMSs. In recent years, Molecular Beam Epitaxy techniques have led to the successful growth of various DMS including $(Ga, Mn)As$. In order to prevent phase separation order, magnetic $Ga_{1-x}Mn_xAs$ should be grown at low temperatures (T; 200–300°C). These problems are not specific to $Ga_{1-x}Mn_xAs$, in fact, all presently known DMS materials suffer from the same problems. As a result, theoretical study of DMS electrical properties such as resistivity is very difficult. Since these properties are influenced by the exchange interaction between the carriers and the localized moments, it must be taken into account non-perturbatively [10, 11]. Due to these influences, spin fluctuation scattering contributes to the resistivity. The experimentally measured dc resistivity in the DMS materials shows interesting behavior strongly depending on the concentration of the magnetic impurity and temperature. In (Ga,Mn)As, there is a metal–insulator transition between insulating samples with small manganese concentration and metallic samples with larger concentration. In and with low concentration (x < 0.03), only an insulating behavior has been observed in transport measurements. Insulating behavior is here characterized by a diverging resistivity for T → 0, indicating localization of carriers. However, near-optimal doping x = 0.05, where the highest value of is reported, the nonmonotonic behavior of insulator–metal–insulator of the resistivity as a function of temperature is observed. A resistivity peak appears near the critical temperature, and the resistivity shows metallic behavior below and insulating behavior at higher temperatures. In metallic samples, the resistivity decreases and eventually saturates for T → 0. The peak has been understood as the critical scattering effects of spin fluctuations [11–13]. In this paper, a Ruderman-Kittel-Kasuya-Yosida (RKKY) interaction was used for the calculation of the resistivity. This paper focuses on the group $\boldsymbol{III - Mn - V}$ ferromagnetic semiconductors, more specifically $Ga_{1-x}Mn_xAs$. Even though we focus on III–V compound based DMS such as $Ga_{1-x}Mn_xAs$, the results presented in this paper are general for all DMS materials.

2. Theoretical formalism

2.1 RKKY interaction approaches to p-d exchange interaction

The carrier-mediated spin–spin coupling is usually described in terms of the Ruderman-Kittel-Kasuya-Yosida (RKKY) model, which provides the energy J_{ij} of exchange coupling. When magnetic impurity Mn is introduced into semiconductors, spin flip processes become possible. The spin of the scattered electron and the spin of magnetic impurities are flipped simultaneously while the component of the total spin along the quantization axis is conserved. The interaction of the impurity with conduction electron is described as p-d or s-d interaction. The spin density of electrons determines this interaction at the localized moment. Like in the Heisenberg model, the scalar product of two spins is taking with the exchange coupling J [14, 15]. Writing the spin density of electrons in terms of the field operators, the interaction between the spin S_j at r_j and the conduction electron is

$$H_{p-d} = -2JS_j.S_i(r_j) = -J\sum\nolimits_{\alpha\beta} S_j.\sigma_{\alpha\beta}\left(\hat{\psi}^+{}_\alpha(r_j)\hat{\psi}_\beta(r_j)\right) \tag{1}$$

Where $\hat{\psi}$ is the field operator states expanded in terms of the block states. Using the field operators, the Fourier transform of number density operator $n(r)$ becomes

$$n(q) = \int n(r)e^{-q.r}dr = \frac{1}{V}\sum\nolimits_{kk'\sigma}\int e^{-i(k+q-k').r}u_k^*(r)u_{k'}(r)c_{k\sigma}^+c_{k'\sigma}dr \tag{2}$$

Where $c_{k\sigma}$ and $c_{k\sigma}^+$ are the annihilation and creation operators of electrons, respectively. The integral in Eq. (2) is vanished unless $k' = k + q + G$ due to the lattice periodicity of the function s $u_k(r)$. Where G is the reciprocal lattice vectors. The coupling strength depends on state k', the state in which the electron of wave vector k is scattered. If we express Eq. (2) in terms of the spin density and the system contains N_i localized spins, the interaction with conduction electrons is determined by

$$H_{p-d} = -\frac{J}{V}\sum_{l=1}^{N_i}\sum\nolimits_{k,k'}\sum\nolimits_{k,k'} e^{i(k-k').r}\left\{S_l^+c_{k'\downarrow}^+c_{k\uparrow} + S_l^-c_{k'\uparrow}^+c_{k\downarrow} + S_l^z\left(c_{k'\uparrow}^+c_{k\uparrow} + c_{k'\downarrow}^+c_{k\downarrow}\right)\right\} \tag{3}$$

The interaction of localized spin Mn ions with electron gives rise to the creation or annihilation of an electron–hole pair. By considering elastic transitions and the common energy $E_k = E_{k'}$ denoted by E_0, the Hamiltonian is written as

$$\langle k'|H_{eff}|k\rangle = -\sum_j \frac{\langle k'|\lambda H_{eff}|j\rangle\langle j|\lambda H_{eff}|k\rangle}{E_j - E_0} \tag{4}$$

Where $|k\rangle$ is ground state, $|k'\rangle$ is the state of the scattering of electron, λ is the coupling constant, and H_{eff} is the effective Hamiltonian, which is equal to H_{p-d}. Since the flip of the impurity spin is accompanied by the flip of an electron spin, the collision term of distribution function of spin-up and spin-down electrons is

$$\begin{aligned}\left(\frac{\partial f(k)}{\partial t}\right)_{coll} &= \sum_{k'} W_{k\uparrow,k'\uparrow}\left\{f_\uparrow(k')\left[1-f_\uparrow(k)\right] - f_\uparrow(k)\left[1-f_\uparrow(k')\right]\right\} \\ &+\sum_{k'} W_{k\uparrow,k'\downarrow}\left\{f_\downarrow(k')\left[1-f_\uparrow(k)\right] - f_\uparrow(k)\left[1-f_\downarrow(k')\right]\right\}\end{aligned} \tag{5}$$

Where $W_{k,k'}$ is the transition probability, $f(k)$ and $f(k')$ are the distribution functions of the initial states and the scattered states, respectively. The transition probability between two states α and β is given by

$$W_{\alpha\beta} = \frac{2\pi}{\hbar}\left|\langle\alpha|H_{p-d}|\beta\rangle\right|^2\delta(E_\alpha - E_\beta) \tag{6}$$

Where $|\alpha\rangle$ is the eigen state of the initial system, and E_α is the corresponding eigen values, $|\beta\rangle$ is the final eigen states, and E_β is the corresponding eigen values, respectively. If the spin is flipped in the intermediate state, an electron of quantum numbers M_l'' appears in the first process, and the impurity spin goes over from the initial state $S^z = M$ to $S^z = M + 1$. Where the electron–hole pair is composed of a spin-up electron and spin-down hole, the impurity spin has to flip down, to the state $S^z = M - 1$. Then by substituting state α and β and using the distribution function for the spin of the electron–hole pair, the total contribution is

$$\langle f|H_{eff}|i\rangle = -\left(\frac{J}{N}\right)^2\sum_{ij}\frac{f_0(E_k)\left[1 - f_0(E_{k'})\right]}{E_k' - E_k}e^{i(k-k')\cdot(r_j-r_i)}\left\{ 2\langle\{M_l'\}|S_i^zS_j^z + \frac{1}{2}\left(S_i^+S_j^- + S_i^-S_j^+\right)|M_l\rangle \right. \tag{7}$$

Where $|\{M''_l\}\rangle$ is the state of the spins of the z-component of the intermediate states, and M_l'' is the corresponding eigen values, $|\{M_l'\}\rangle$ is the state of the spins of the z-component of the scattered states, and M_l' is the corresponding eigen value, and $|\{M_l\}\rangle$ is the state of the spins of initial state, and M_l is the corresponding eigen value.

Eq. (7) can be considered as the matrix element of the operator H, when we write the exchange energy between S_i and S_j, which is written as

$$H = -\sum_{ij} J(r_i - r_j)S_i.S_j \tag{8}$$

To determine the exchange coupling strength $J(r_i - r_j)$, we introduce $\boldsymbol{r} = \boldsymbol{r_i} - \boldsymbol{r_j}$ where $\mathbf{r}$ is the displacement vector between the ions i and j, and using the following integral notation I as

$$I = (1/V)^2\sum_{kk'}\frac{f_0(E_k)\left[1 - f_0(E_{k'})\right]}{E_k' - E_k}e^{i(k-k').r} \tag{9}$$

Replacing the sum in Eq. (9) by an integral, using the quadratic dispersion relation valid for free electrons and the notations $\kappa = kr, \kappa' = k'r$, the parabolic energy $E_k = \frac{\hbar^2k^2}{2m^*}$, in which m^* is the effective mass, and trigonometric identity $sin\,\kappa' = \frac{e^{i\kappa'} - e^{-i\kappa'}}{2i}$ and the complex variable $\kappa' + i\eta$ instead of κ', where η is an infinitesimal quantity, the angular integral becomes

$$I = -\frac{m_e k_{FJ^2}^4}{\hbar^2\pi^3}\frac{sin\,2k_Fr - 2k_Frcos2k_Fr}{(2k_Fr)^4} \tag{10}$$

Then the effective exchange interaction or the coupling exchange interaction constant between magnetic impurities and delocalized charge carriers is given by

$$J(r) = \frac{m_eJ^2k_F^4}{\hbar^2\pi^3}F(2k_Fr) \tag{11}$$

Where k_F is the Fermi wave vector and the function $F(x) = \frac{x\cos x - \sin x}{x}$, $x = 2k_F r$. Eq. (11) gives the well-known RKKY interaction. As the value of the oscillatory function $F(2k_F r)$ varies, the value of the exchange interaction is either positive or negative, and hence, the interaction in the system can be ferromagnetic or antiferromagnetic.

By using Eq. (11), we can determine the relaxation time τ of the impurity spin of electrons. For spin conserving scattering, the matrix element of Eqs. (7) and (8) is

$$\langle k'|H_{eff}|k\rangle = -\frac{J}{V}S^z \tag{12}$$

A better approximation can be obtained for the transition probability by replacing the matrix element of the scattering matrix T in Eq. (6). Then Eq. (6) is rewritten as

$$W_{\alpha\beta} = \frac{2\pi}{\hbar}|\langle\alpha|T|\beta\rangle|^2\delta(E_\alpha - E_\beta) \tag{13}$$

The scattering (transition) probability in Eq. (13) can be expressed in terms of the interaction Hamiltonian up to third order in the coupling constant. Then using the second-order correction to the T matrix, the Fermi-Dirac distribution function, which is expressed by $f_0(k) = \langle c_k^+ c_k\rangle$ and if the z-component of the impurity spin is unchanged by the scattering, the combined contribution is

$$\left(-\frac{J}{V}\right)^2 \sum_{k''} (S^z)^2 \frac{(1 - f_0(k''))}{E_k - E_{k''}} - \frac{f_0(k'')}{E_{k''} - E_{k'}} = \left(-\frac{J}{N}\right)^2 (S^z)^2 \sum_{k''} \frac{1}{E_k - E_{k'}} \tag{14}$$

If the impurity spin goes over from the initial state $S^z = M$ to $S^z = M + 1$, the electron–hole pair is composed of a spin-up electron and a spin-down hole (i.e., the impurity spin has to flip down) to the state $S^z = M$ to $S^z = M - 1$, and on the account of conservation of energy $E_k = E_{k'}$, then the combined contribution of these processes gives

$$\left(-\frac{J}{V}\right)^2 \sum_{k''} \left\{ \left[S(S+1) - M^2\right] \frac{1}{E_k - E_{k''}} - M\frac{1 - 2f_0(k'')}{E_k - E_{k''}} \right\} \tag{15}$$

The first term in Eq. (15) is negligible since it is small; however, the second term is no longer small. The second term diverges logarithmically. Using this form, the transition probability Eq. (13) in J by summing over the possible spin orientations is given by

$$W(k\uparrow, k'\uparrow) = W(k\uparrow, k'\downarrow) = \frac{N_i 2\pi J^2 S(S+1)}{3\hbar}[1 + 4Jg(E_k)]\delta(E_k - E_{k''}) \tag{16}$$

Where N_i is the concentration of magnetic impurities, which is $N_i = 4x/a_{lc}^3$ where x is the concentrations of the impurity added to the semiconductors, a_{lc} is lattice constant, and the singular function in the third order correction is

$$g(E_k) = \frac{1}{V}\sum_{k''}\frac{f_0(k'')}{E_k - E_{k''}} \tag{17}$$

We shall investigate the dependence of Eq. (16) on the energy of the initial state E_K, which is entirely involved in the function $g(E_k)$. At the absolute zero of temperature, $f_0(k'')$ can be replaced by a step function, which is unity when $k'' < k_0$ and zero when $k'' > k_0$, where k_0 is the magnitude of the Fermi momentum.

By using Fermi-Dirac distribution, $E_k = \frac{\hbar^2 k^2}{2m^*}$ and for electron close to the Fermi surface at low temperature, the singular function becomes

$$g(E_k) = \left(\frac{3z}{2E_F}\right)\{1 + (k/2k_0)\, log\, |(k - k_0)/(k - k_0)|\} \tag{18}$$

Where z is the number of conduction electrons per atom. Since at T $\neq 0$, the average of $|k - k_0|$ for thermally excited electrons is proportional to T, even at this stage of calculation we can expect a term proportional to $\log T$ in the expression of the resistivity. It is true that $g(E_k)$ diverges when E_k approaches to E_F. Eq. (18) makes the calculation easier to retain the definition of $g(E_k)$ here and to carry out the summation after we get the expression for the resistivity.

Then substituting Eq. (16) into collision integral, the relaxation time τ is the inverse of the transition probability. Therefore, the spin-dependent relaxation time is

$$\frac{1}{\tau} = \frac{N_i 2\pi J^2 S(S+1)}{3\hbar E_F}[1 + 4Jg(E_k)] \tag{19}$$

The Fermi energy E_F is given by $E_F = \frac{\hbar^2 k^2}{2m^*} = \frac{\hbar^2}{2m^*}\left(3\pi^2\left(\frac{N}{V}\right)\right)^{\frac{2}{3}}$. In $III - V$ semiconductors, it is found that the maximum and minimum energy locations of the Fermi level typically do not deviate by more than 1ev. In the specific case of $GaAs$, the conduction band is located at $E_{FS} + 0.9eV$ and the valence band at $E_{FS} - 0.5eV$ [3]. Where E_{FS} is the Fermi-level stabilization energy, and it is found to be located at $\sim 4.9eV$ below the vacuum level and is the same for all III-V and II-VI SCs. since Mn-doped $GaAs$ is a p-type, we use Fermi level energy of the valence band. By using this value of Fermi energy and the values of other constant, we can calculate τ_0.

In the presence of external perturbation due to the electric field or temperature gradient, there is a variation in the function $f_k^{\pm}$. Supposing the electric field lies along the z-direction, then using Eq. (19), the rate change of the probability $f_k^{\pm}$ with which the state $K_{\pm}$ is occupied due to the collision with the localized spins becomes

$$\left(\frac{\partial f_k^{\pm}}{\partial t}\right)_{coll} = -\frac{(f_k^{\pm} - f_k^0)}{\tau} \tag{20}$$

Then from standard theory, we can easily obtain the resistivity as

$$\rho_{spin} = x\rho_0\left\{1 - \left(\frac{(\hbar^2 j}{\pi m_h{}^* k_0}\right)\int g(E_k)\left(\frac{df^0}{dE_k}\right)d^3k\right\} \tag{21}$$

Where ρ_0 is defined by

$$\rho_0 = 3\pi m_h{}^* J^2 S(S+1)(V/N)/2e^2\hbar E_F \tag{22}$$

Where V is the volume of the crystal. Then using some techniques, the total electrical resistivity becomes

$$\rho_{spin} = x\rho_0\{1 - (3zJ/E_F)\log T\} \tag{23}$$

Since substitutional Mn ions in $GaAs$ (i.e., $Ga_{1-x}Mn_xAs$) are a p-type semiconductor, we use the hole concentration $p = \frac{N}{V}$, instead of n. The hole concentration p would ideally to be given by $p = xN_0$ where $N_0 = 2.2x10^{22}cm^{-3}$ is the concentration of Ga sites in $GaAs$ and x is the concentration of impurity [3]. For $(Ga, Mn)As$, the hole effective mass $m_h^* = 0.5m_0$, where $m_0 = m_e$ is the electron mass [16].

3. Results and discussion

As mentioned in Section 2, the temperature dependence of electrical resistivity is calculated. The electrical resistivity of p-type Mn-doped $GaAs$ has been investigated theoretically. This theoretical calculation is done using RKKY interaction models. This theoretical calculation is somewhat similar to experimental work, which is done on this material. In this paper, we use RKKY to calculate the exchange interaction constant and spin-dependent relaxation time approximation. Using these exchange interaction constant and spin-dependent relaxation time, the electrical resistivity of this material is calculated.

Electrical resistivity is the inverse of electrical conductivity of the material. The electrical resistivity of semiconductors decreases exponentially with increasing temperature in contrast to that of pure metals. The electrical resistivity of extrinsic semiconductor is decreased with increasing both the concentration of magnetic impurity and the temperature. From this concept the electrical resistivity of DMSs, in

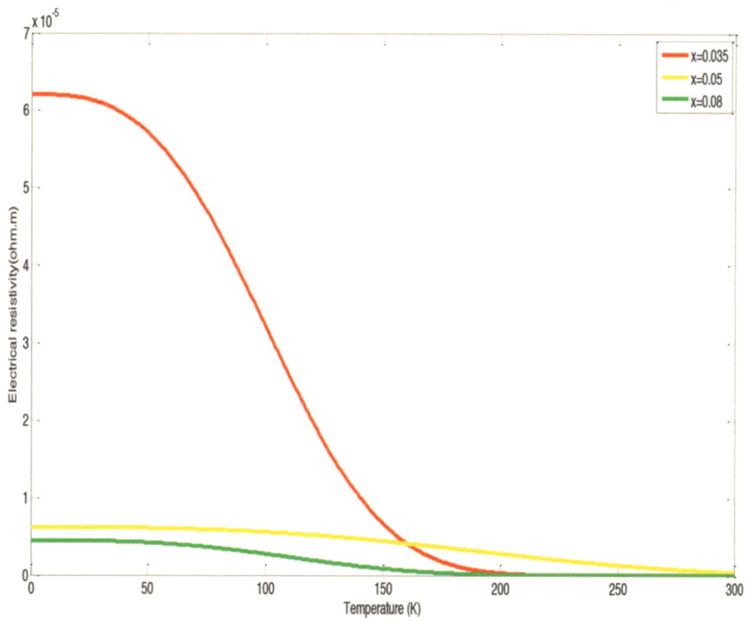

Figure 1.
Electrical resistivity of $Ga_{0.92}Mn_{0.08}As$ versus temperature.

which magnetic impurities are incorporated into standard semiconductors, is decreased with temperature and concentration of magnetic ions. The temperature dependence of resistivity in $Ga_{1-x}Mn_xAs$ for $0.08 \leq x \geq 0.01$ is shown in **Figure 1**. From Eq. (23) the electrical resistivity is inversely proportional to the hole density or hole concentration (i.e., $\rho \propto \frac{1}{p}$). Since $Ga_{1-x}Mn_xAs$ is an extrinsic semiconductor, its hole concentration is increased with adding the impurity. Then the electrical resistivity decreases with increasing concentration. From **Figure 1** the electrical resistivity of $Ga_{1-x}Mn_xAs$ for x = 0.035, x = 0.05, and x = 0.08 is drawn. From the graph as the magnetic impurity concentration is increased, its resistivity decreases. As the temperature and hole concentration increase, the electrical resistivity is decreased. At x = 0.08 or at high concentration, there is low resistivity and at x = 0.035, the resistivity is high.

4. Conclusion

In this paper, we studied the theoretical temperature dependence of electrical resistivity of DMS specifically $Ga_{1-x}Mn_xAs.Ga_{1-x}Mn_xAs$ has become the most understood and extensively studied $III_{1-x}Mn_xV$ ferromagnetic semiconductor. DMSs show interesting electrical properties, which strongly depend on the concentration of magnetic dopants and temperature. These properties of $Ga_{1-x}Mn_xAs$ is affected by concentration and temperature. Since the electrical properties are dependent on the density of states, the impurity added to the semiconductor increases this density of state and decreases energy gap. The electrical resistivity of $Ga_{1-x}Mn_xAs$ is exponentially decreased with an increase of temperature from $0K - 300K$ and concentration of magnetic ions in the range of $0.08 \leq x \geq 0.01$. In general, if T_C of the DMSs can be increased, there is a possibility of utilizing the system under consideration for the spintronic purpose at room temperature.

Acknowledgements

First of all, I would like to thank the almighty God for keeping me safe and continually blessing me in all aspects of my life. I extend my heartful thanks to my parents for their encouragement and support both morally and psychologically.

Conflicts of interest

The authors declare no conflicts of interest.

Data availability

No data were used to support this study.

Author details

Edosa Tasisa Jira
Department of Physics, College of Natural and Computational Sciences, Mekdela Amba University, Dessie, Ethiopia

*Address all correspondence to: edosatasisa17@gmail.com

References

[1] Dietl T. Ferromagnetic semiconductors. Semiconductor Science and Technology. 2002;**17**:377-392

[2] Sapoval B, Hermann C. Physics of Semiconductors. New York: Springer; 1995. pp. 1-2

[3] Dietl T, Awschalom DD, Kaminska M, Ohno H. Semiconductor and Semimetals. Vol. 82. Elseiver; 2008

[4] Shinjo T. Nanomagnetism and Spintronics. 1st ed. Elsevier; 2009. pp. 20-22

[5] Zheng Y. A prespective of recent progress in *ZnO* DMSs, Springer-Verlag Berlin Heidelberg. Applied Physics A. 2013;**112**:241–254

[6] Twardowski A. Diluted magnetic *III-V* semiconductors, international school of semiconducting compounds. 2000;**98**

[7] Kossut J, Gaj JA. Introduction to the Physics of Diluted Magnetic Semiconductors. Berlin Heidelberg: Springer-Verlag; 2010. p. 356

[8] Macdonald AH, Schiffer P, Samarth N. Ferromagnetic Semiconductors. Nature Publishing Group; Nature Materials. 2005;**4**

[9] Chen L, Yang X, et al. Enhancing the curie temperature of ferromagnetic semiconductor (Ga,Mn)As to 200 K via nanostructure engineering. Nano Letters. 2011;**11**:2584-2589

[10] Kaminski A, Das Sarma S. Magnetic and transport percolation in diluted magneticSemiconductors. Physical Review B. 2003;**68**:235210

[11] Hwang EH, Das Sarma S. Transport properties of diluted magnetic semiconductors. Physical Review B. 2005;**72**:035210

[12] Matsukura F, Ohno H, Shen A, Sugawara Y. Physical Review B. 1998;**57**: R2037

[13] Timm C. Disorder effects in diluted magnetic semiconductors. Journal of Physics: Condensed Matter. 2003;**15**: R1865

[14] Dietl T. Ferromagnetic transition in diluted magnetic semiconductors. Condensed Matter Physics. 1999;**2**(3 (19)):495-508

[15] Solyum J. Fundamentals of the Physics of Solids. Vol. II. Berlin Heidelberg: Springer-Verlag; 2009. pp. 213-218

[16] Blakemore JS. Semiconducting and other major properties of gallium arsenide. Journal of Applied Physics. 1982;**53**(10)

Chapter 6

Radiation Response of Group-IV and III-V Semiconductors Subjected to D–D and D–T Fusion Neutrons

Jean-Luc Autran and Daniela Munteanu

Abstract

This work focuses on the radiation response of Group IV (Si, Ge, SiC, diamond) and III-V (GaAs, GaN, GaP, GaSb, InAs, InP, InSb, AlAs) semiconductors subjected to D–D (2.45 MeV) and D–T (14 MeV) neutrons. The response of each material has been systematically investigated through a direct calculation using nuclear cross-section libraries, MCNP6, and Geant4 numerical simulations. For the semiconductor materials considered, we have investigated in detail the reaction rates per type of reaction (elastic, inelastic, and nonelastic) and proposed an exhaustive classification and counting of all the neutron-induced events and secondary products as a function of their nature and energy. Several metrics for quantifying the susceptibility of the related semiconductor-based electronics to neutron fusions have been finally considered and discussed.

Keywords: radiation effects, III-V compound semiconductors, fusion neutrons, D-D reaction, D-T reaction, numerical simulation, neutron cross section, elastic scattering, inelastic scattering, nonelastic interactions, nuclear data library, Geant4, MCNP, single-event effect

1. Introduction

At the horizon of 2035 and beyond, the very large electronic equipment for command and diagnostic operations of future fusion power machines, such as ITER, the international thermonuclear experimental reactor [1], will embrace a wide variety of semiconductors materials. Indeed, new materials, such as SiC, Ge, GaN, or GaAs, are envisaged to progressively replace the traditional silicon of microelectronics for high temperature, high performance, or high-speed electronics/optoelectronics applications [2–5]. These equipment will be directly or partially (behind radiation shielding) exposed to a complex radiation field due to primary fusion neutrons created in the core of the plasma and then transported outside the machine [6]. These neutrons are produced in the deuterium–tritium (D–T) reaction that has been identified as the most efficient for fusion devices [7]:

 IntechOpen

$$D + T \rightarrow {}^{4}He\ (3.52MeV) + n\ (14.07MeV) \quad (1)$$

However, one must also consider deuterium–deuterium (D–D) reaction because this less efficient fusion reaction will be used to study, during the development and commissioning of future machines, their operation before introducing tritium fuel. D–D fusion following the two main reactions with equal probability of occurrence:

$$D + D \rightarrow {}^{3}He\ (0.82MeV) + n(2.45MeV) \quad (2)$$

$$D + D \rightarrow T\ (1.01MeV) + p\ (3.02MeV) \quad (3)$$

For future fusion power machines, the radiative reliability issue for the envisaged new material-based electronics arises now, even if—(i) most of these components have not yet been manufactured and tested, and (ii) the design and the integration of machine electronics equipment will not take place for many years. In such a long-term development, numerical simulation offers a very promising solution to anticipate, in the first step at the material level, this reliability issue for semiconductors whose radiative response is little or poorly known.

In the direct continuation of our previous works [8–10] that investigated the atmospheric radiation response of a wide variety of semiconductors, the present chapter focuses on the radiation response of Group IV (Si, Ge, SiC, diamond) and III-V (GaAs, GaN, GaP, GaSb, InAs, InP, InSb, AlAs) semiconductors subjected to D–D (2.45 MeV) and D–T (14 MeV) neutrons.

In this work, the response of each material has been systematically investigated through a direct calculation using nuclear cross-section libraries, MCNP6, and Geant4 numerical simulations. In the following, for the 12 semiconductor materials considered, we will investigate in detail the reaction rates per type of reaction (elastic, inelastic, and nonelastic) and propose an exhaustive classification of all the neutron-induced secondary products as a function of their nature and distribution in the energy. Implications for quantifying the susceptibility of these related semiconductor-based electronics to neutron fusions will be finally presented and discussed.

2. Material properties

Table 1 summarizes the natural isotopic abundance of the different chemical elements entering the composition of the studied semiconductor materials. Nitrogen, phosphorus, and arsenic have only a single natural isotope which simplifies the calculations to come for these elements. On the contrary, the other elements have two (C, N, Ga, In, Sb), three (Si), and up to five (Ge) different natural isotopes. Nevertheless, for C and N, the relative weight of the minority isotope is less than or around the percent, so we can consider that the natural carbon and nitrogen consist only of the simple isotopes ^{12}C and ^{14}N, respectively, with a very good approximation. The other isotopes listed in **Table 1** must be considered with their respective weight in atomic composition for correctly simulating the neutron response of the natural materials to which they relate.

Table 2 reports the value of the energy bandgap, the number of atoms per square centimeter, the density, and the value of the energy for electron-hole pair creation in bulk material for the 12 semiconductors studied in this work. Due to their atomic composition and crystallographic structure, all these materials are denser than Si (2.33 g/cm^3) and three materials exhibit a much larger number of atoms per square centimeter than Si—SiC, diamond, and GaN. This can have a direct consequence on the

Symbol	Z	Nuclide (100×Z+A format)	Natural abundance (%)
C	6	6012	98.93
		6013	1.07
N	7	7014	99.6
		7015	0.4
Al	13	13027	100
		14028	92.22
Si	14	14029	4.68
		14030	3.09
P	15	15031	100
Ga	31	31069	60.1
		31071	39.9
		32070	20.52
		32072	27.45
Ge	32	32073	7.76
		32074	36.52
		32076	7.75
As	33	33075	100
In	49	49115	95.7
		49113	4.3
Sb	51	51121	57.4
		51123	42.6

Table 1.
Natural abundance of nuclides related to the materials studied in this work. Z refers to the atomic number, A to the atomic mass.

Material	Bandgap at 300 K (eV)	Number of atoms per cm^3 ($\times 10^{22}$)	Density (g/cm^3)	E-h pair creation energy Eeh (eV)
Si	1.124	5.0	2.329	3.6
Ge	0.661	4.42	5.3267	2.9
SiC (4H)	3.23	9.64	3.21	7.8
Diamond	5.47	17.6	3.515	12
GaAs	1.42	4.42	5.32	4.8
AlAs	2.16	4.42	3.76	6.8
InP	1.34	3.96	4.81	4.5
InAs	0.36	3.59	5.67	1.8
GaSb	0.73	3.53	5.61	2.7
InSb	0.17	2.94	5.78	1.1
GaN	3.39	8.85	6.15	8.9
GaP	2.26	4.94	4.138	6.8

Table 2.
Main properties of the semiconductors considered in this study. From Ref. [8, 10].

number of neutron interactions per unit volume of the material, as discussed in the following. In addition, seven materials (SiC, diamond, GaAs, AlAs, InP, GaN, and GaP) exhibit a bandgap larger than Si and logically, a larger electron-hole pair creation energy (3.6 eV for Si) [10], four materials (Ge, InAs, GaSb, and InSb) are low-bandgap semiconductors (< 1eV) with lower electron-hole pair creation energy than Si. The electrical consequences of this important parameter are discussed in Section 5.

3. Direct evaluation of semiconductor susceptibility to fusion neutrons

The susceptibility of the different semiconductors to neutron irradiation can be firstly evaluated via the calculation of the number of neutron-material interactions under 2.45 or 14 MeV neutrons. This can be performed via a direct analytical calculation using neutron cross-section library data; we considered here the TENDL-2021 nuclear data library [11]. The number of interactions in the target at a fixed neutron energy E is simply given by:

$$N(E) = \sum_i f_i \, \sigma_i(E) \times 10^{-24} \times NV \times e \times M \tag{4}$$

where σ_i is the value at energy E of the cross section for isotope i (in barn), fi is the fraction of isotope i in the target isotopic composition, e is the target thickness, NV is the number of atoms per unit volume, and M is the number of incident monoenergetic neutrons; in this work, fixed to the arbitrary value of 5×10^8 for standardization purpose (introduced in other previous studies [8–10], M initially corresponds to the number of atmospheric neutrons impacting a surface of 1 cm^2 at sea level exposed to natural radiation during 25×10^6 h).

Figure 1 illustrates the extraction from TENDL-2021 library data of the neutron cross-section values at 2.45 and 14 MeV for two different isotopes. Total, elastic, and inelastic data, respectively, correspond to reaction type numbers MT = 1, 2, and 4 in the standardized ENDF format [12]. When it is not available, the cross-section value for the sum of all nonelastic channels (MT = 3) is obtained by subtracting elastic + inelastic cross sections to the total cross section.

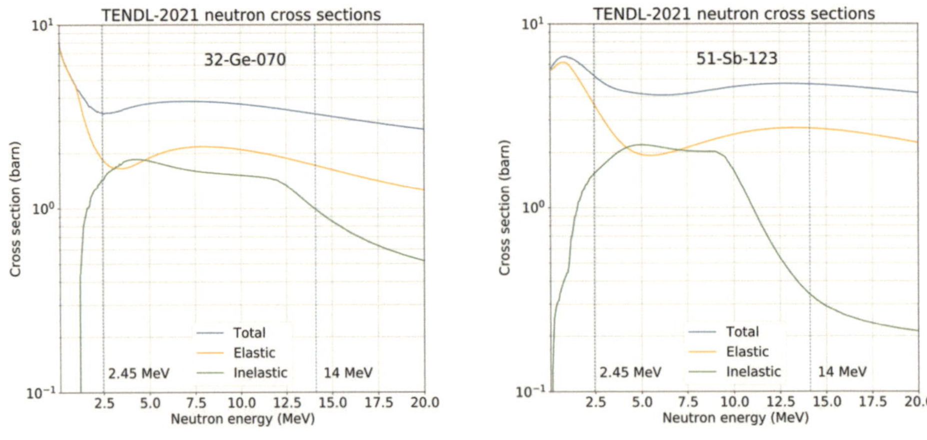

Figure 1.
Example of neutron cross sections (total, elastic and inelastic) extracted from TENDL-2021 nuclear data library [11] for the isotopes 32-Ge-070 (left) and 51-Sb-123 (right).

As the result of this analytical evaluation using Eq. (4), **Table 3** gives the total number of interaction events in a target (1cm^2 × 20 μm) composed of a single chemical element with an isotope distribution corresponding to the natural abundance (**Table 1**) and subjected to 2.45 or 14 MeV neutrons. Values reported in **Table 3** correspond to a normalized value NV = 1.0 × 10^{22} at/cm^3 for all targets.

From the data of **Table 3**, it is now easy to evaluate the neutron response for any material composed of these elements. **Figure 2** shows the results of this evaluation for the different semiconductor materials, considering their stoichiometry and their exact number of atoms per square centimeter, given in **Table 2**. All numerical values for **Figure 2** are detailed in the Appendix (**Table 6**). From the results of **Figure 2**, we can formulate the following observations:

- Elastic and inelastic interactions dominate for all materials at 2.45 MeV whereas nonelastic interactions represent very roughly one-third of the total number of events at 14 MeV.

- All studied materials exhibit a larger total number of interaction events than silicon, both at 2.45 and 14 MeV.

- Two semiconductors, C (diamond) and GaN show the largest neutron susceptibility at these two energies, followed by SiC.

- Diamond at 2.45 MeV exhibits a pure and extremely high elastic response; this behavior is due to the combination of an important elastic cross section (1.62 barn at 2.45 MeV) and the highest NV value (1.76 × 10^{23} at/cm^3) between all materials.

- Only a single material, GaN, demonstrates a frank higher response at 14 MeV than at 2.45 MeV in terms of total events (elastic + inelastic + nonelastic),

Target (1.0×10^{22} at/cm^3)	Z	N(E) from Eq.(4) and TENDL-2021					
		2.45 MeV			14 MeV		
		Elastic	Inelastic	Nonelastic	Elastic	Inelastic	Nonelastic
C	6	16,146	0	0	8360	4240	674
N	7	13,201	22	1675	9697	3658	2328
Al	13	25,564	4171	61	7798	4598	5050
Si	14	18,857	6321	0	7555	4922	6706
P	15	28,196	5378	472	7522	2910	8552
Ga	31	16,150	15,009	61	16,839	5908	9892
Ge	32	17,572	16,013	329	17,203	7306	8626
As	33	16,008	17,718	649	17,934	5874	10,119
In	49	31,349	22,255	1735	25,669	4173	16,178
Sb	51	35,376	15,712	612	26,909	3832	15,941

Table 3.
*Total number of interaction events in a target (1cm^2 × 20 μm) composed of a single chemical element with an isotope distribution corresponding to the natural abundance (**Table 1**) and subjected to 2.45 or 14 MeV neutrons. Values are calculated from Eq. (4) with a normalized value NV = 1.0 × 10^{22} at/cm^3 for all targets.*

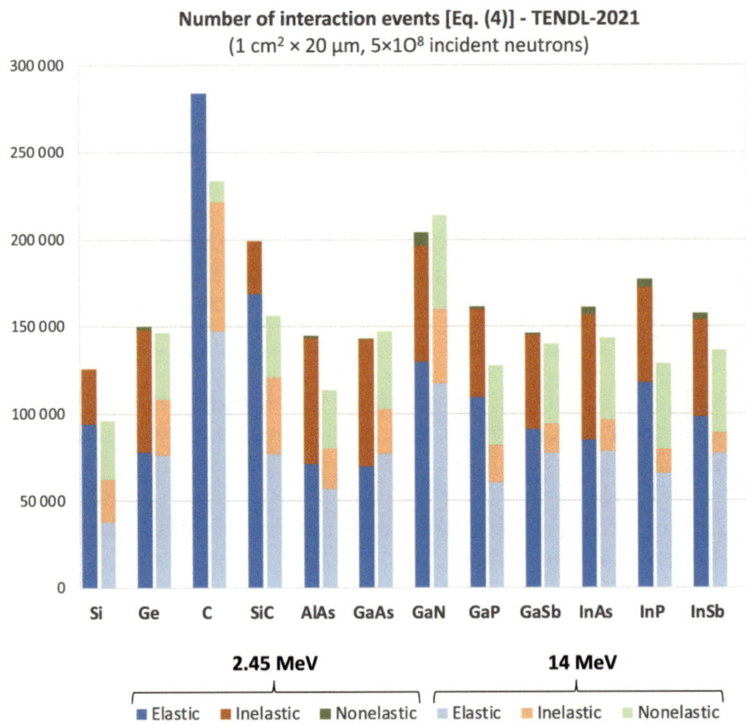

Figure 2.
Number of elastic, inelastic, and nonelastic interaction events in a target ($1cm^2 \times 20$ µm) composed of natural semiconductors and subjected to 5×10^8 neutrons of 2.45 or 14 MeV. Values are deduced from ***Table 3*** *considering the stoichiometry and the exact number of atoms per square centimeter for each material given in* ***Table 2****. Numerical values for this graph are reported in the appendix (****Table 6****).*

due to an elevated number of nonelastic events, the highest at 14 MeV for all materials.

- AlAs exhibits the smallest neutron response after silicon.

All these results are now investigated in detail in the next section via more complete numerical simulations.

4. Geant4 and MCNP6 simulations

4.1 Simulation details

The neutron susceptibility of the studied materials has been in-depth analyzed using the Monte Carlo radiation transport codes Geant4 (version 10.07 patch 02, G4NDL4.6 neutron library [13–15]) and MCNP6 (version 6.2 [16, 17]). In a similar way to previous analytical estimations, natural semiconductor targets ($1 \text{ cm}^2 \times 20$ µm) have been subjected to incoming 2.45 or 14 MeV neutrons perpendicularly to their largest surface. The same number of 5×10^8 incident neutrons has been considered for each simulation run.

For Geant4, a simulation run produces a single output file containing all the information related to the neutron interaction events in the target material—nature of the interaction, spatial coordinates of the reaction vertex, and exhaustive list of secondary particles produced during the interaction (energy and emission direction vector for each of these emitted particles). A post-treatment eliminates all gamma photons, neutral and light particles (e^-, e^+, η), not able to induce significant single-event effects in electronics.

For MCNP6 simulation, PTRAC card options were activated to obtain, for each event, the nature and the coordinates of the vertex of the interaction and the final energy of the neutron after the interaction. It is thus possible to count and to discriminate interactions as a function of the target atom and as a function of their nature (elastic, inelastic, and nonelastic).

4.2 Numbers of interaction events at 2.45 and 14 MeV

Figures 3 and **4** show the number of interaction events for the different semiconductor targets subjected to 2.45 and 14 MeV neutrons, respectively. All the numerical values for these two figures are reported in **Table 6** of the Appendix. These results show a good agreement, on one hand, between analytical and numerical estimations, and, on the other hand, between Geant4 and MCNP6 results. If we consider the total number of interaction events (**Table 6**), analytical and numerical results agree within 6% on average at 2.45 MeV and within 2.4% on average at 14 MeV. Between Geant4 and MCNP6 results, agreements are better, within 3% on average at

2.45 MeV and within 1.6% on average at 14 MeV. The best agreement is obtained for diamond, GaN, and GaAs at 14 MeV, the largest difference is for GaP at 2.45 MeV,

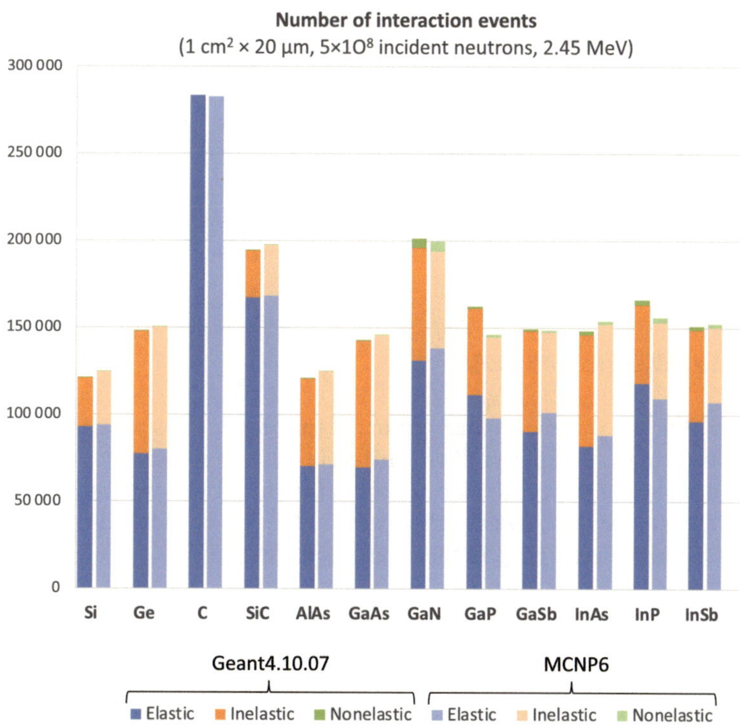

Figure 3.
Number of elastic, inelastic, and nonelastic events estimated by Geant4 and MCNP6 as a function of the nature of the target material for 2.45 MeV incoming neutrons.

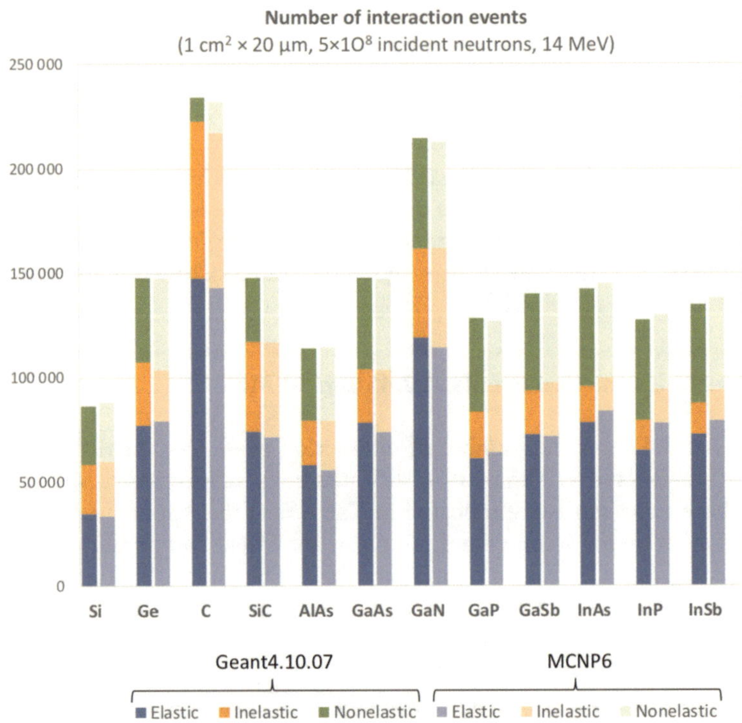

Figure 4.
Number of elastic, inelastic, and nonelastic events estimated by Geant4 and MCNP6 as a function of the nature of the target material for 14 MeV incoming neutrons.

with 11.5% of the variation between Geant4 and MCNP6, the number of elastic events being larger than 14% for Geant4 with respect to MCNP6. Globally, one can see that the different sets of results agree very satisfactorily.

In the following and because Geant4 can track all particles and, consequently, give more exhaustive information about secondaries than MCNP6, we will explore in detail the distributions and characteristics of secondaries produced in the interaction events at both 2.45 and 14 MeV from Geant4 results.

4.3 Detailed analysis at 2.45 MeV

Figures 5–7 shows the energy histograms of the secondaries produced by 2.45 MeV neutrons in the targets composed of the different natural semiconductor materials under investigation. In the particular case of C (diamond), shown in **Figure 5**, all secondaries are exclusively elastic carbon recoil nuclei. They form a continuum from zero to a maximum energy E_{max} theoretically equal to [18]:

$$E_{max} = 4 \times E \times \frac{A}{(A+1)^2} \tag{5}$$

where A is the mass number of the target nucleus and E is the energy of the incident neutrons. Here, for carbon, A = 12 and with E = 2.45 MeV, we obtain a theoretical value of E_{max} = 0.696 MeV, in excellent agreement with the cutoff value of 0.7 MeV extracted from **Figure 5**. This figure also shows a direct comparison between

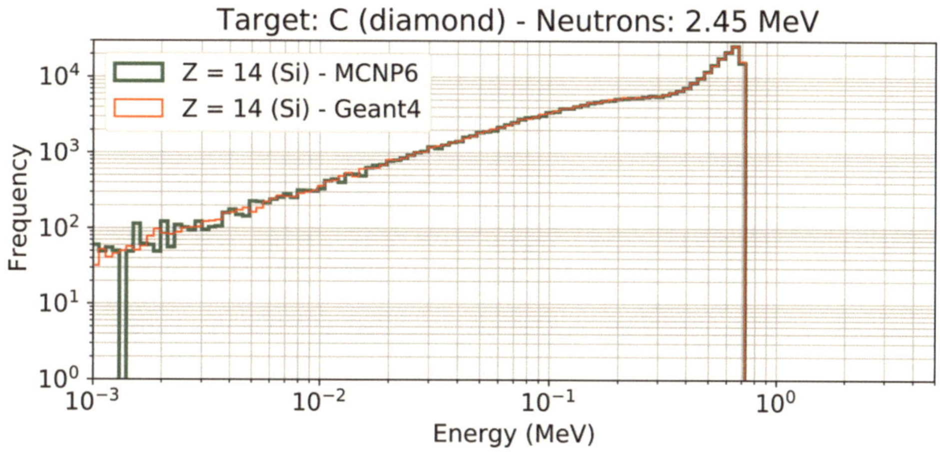

Figure 5.
Energy histograms of secondaries produced in the diamond target by 2.45 MeV neutrons and obtained from Geant4 and MCNP6 simulations. All secondaries are elastic recoil nuclei.

MCNP6 and Geant4 simulation results because, in the case of elastic events, it is possible from MCNP6 output data to compute the energy distribution of recoil nuclei; an estimation unfortunately not possible for inelastic and nonelastic events.

Figures 6 and 7 show Geant4 results for the other target materials. For Si, Ge, SiC, AlAs, GaAs, InAs, InSb, and GaSb (**Figure 6**), 2.45 MeV neutrons produce quasi exclusively, via elastic, inelastic, and nonelastic interactions, secondaries with the same atomic number than one of the elements of the target. The heavier the recoil nucleus or secondary product, the more the distribution is shifted toward the lower energies. This shift is more marked as the difference in atomic number is greater, for example for InP (Z = 15 and 49). Only three materials, i.e., GaN, GaP, and InP, show (**Figure 7**) a significant production of secondaries at 2.45 MeV that differ from target elements:

- For GaP and InP, the presence of phosphorus makes possible the nonelastic reaction $^{31}P(n,p)^{31}Si$ that produces a proton with mean energy around 890 keV. Following this reaction scheme, 980 protons are emitted in GaP and 755 in InP. $^{31}P(n,p)^{31}Si$ represents 88% (resp. 28%) of the nonelastic interactions in GaP (resp. in InP).

- For GaN, the presence of nitrogen makes possible the two nonelastic reactions $^{14}N(n,\alpha)^{11}B$ and $^{14}N(n,p)^{14}C$ that respectively produce an alpha particle (around 1.6 MeV) and a proton (around 2.7 MeV). 4555 alphas and 614 protons are emitted following these two reaction schemas. $^{14}N(n,\alpha)^{11}B$ and $^{14}N(n,p)^{14}C$, respectively, represent 84 and 11% of the nonelastic interactions in GaN.

In addition to the previous analysis, **Figure 8** (left) gives, for each material, the sum of the initial kinetic energy for all produced secondaries (we recall here that neutrons, gamma rays, and light particles, such as electrons or positrons are not considered, only charged particles with $Z \geq 1$ are considered). Such a quantity represents the energy susceptible to be deposited by all these secondaries in the considered target if they are totally stopped (after transferring the totality of their kinetic energy, essentially by ionization process). This energy amount is the highest for C (about 10^5 MeV), followed by SiC, GaN, Si, and GaP (all above 10^4 MeV); it is minimum for InSb (around 3×10^3

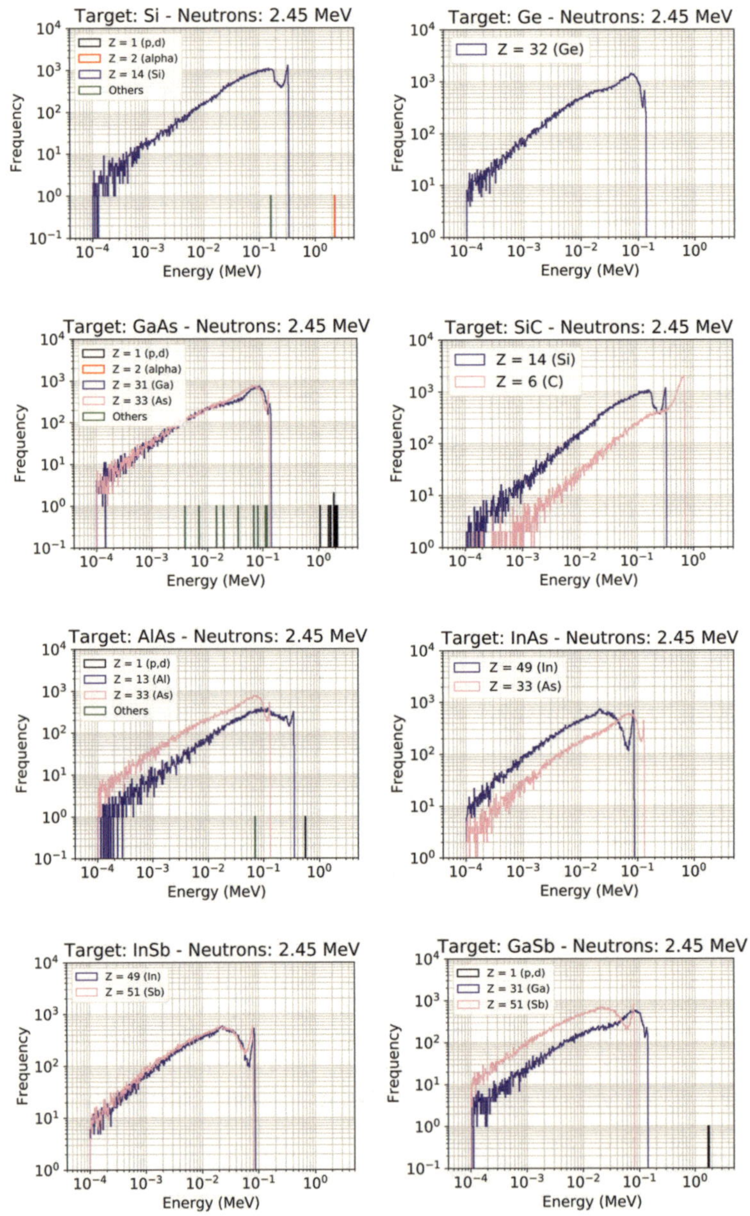

Figure 6.
Energy histograms of secondaries produced by 2.45 MeV neutrons in targets of Si, Ge, GaAs, SiC, AlAs, InAs, InSb, and GaSb semiconductor materials.

MeV). **Figure 8** (right) gives the average energy conveyed per particle for the four categories of secondaries. The consequence of this result on the electrical point-of-view and in the framework of single-event effect occurrence is be discussed in Section 5.

4.4 Detailed analysis at 14 MeV

Figures 9 and **10** show the energy histograms of the secondaries produced by 14 MeV neutrons in the different targets. Contrary to the previous case at 2.45 MeV, a

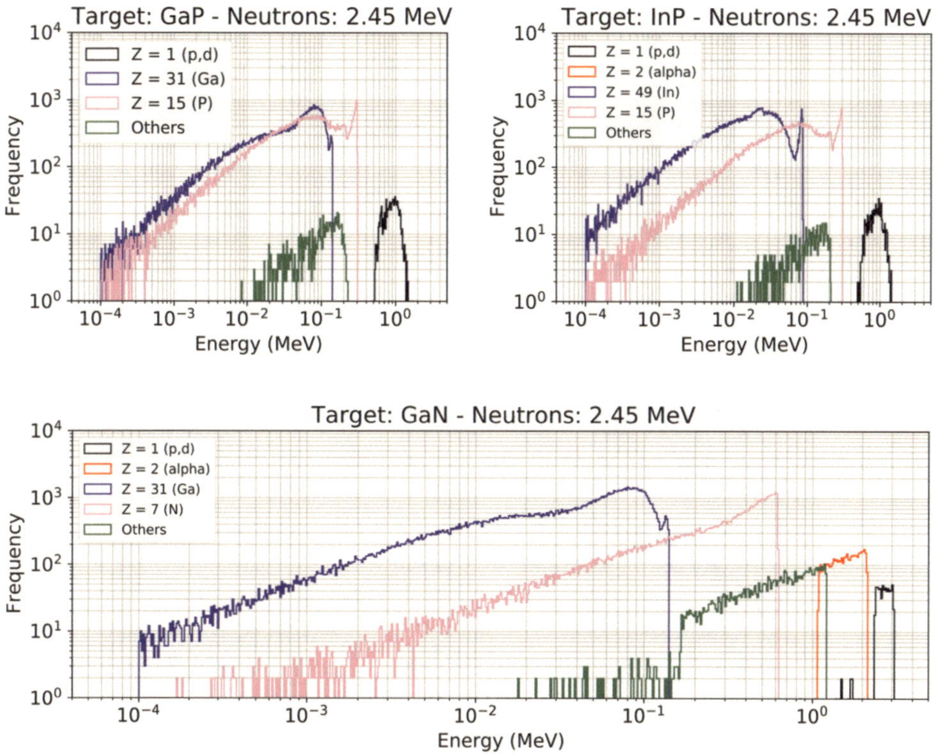

Figure 7.
Energy histograms of secondaries produced by 2.45 MeV neutrons in targets of GaP, InP, and GaN semiconductor materials.

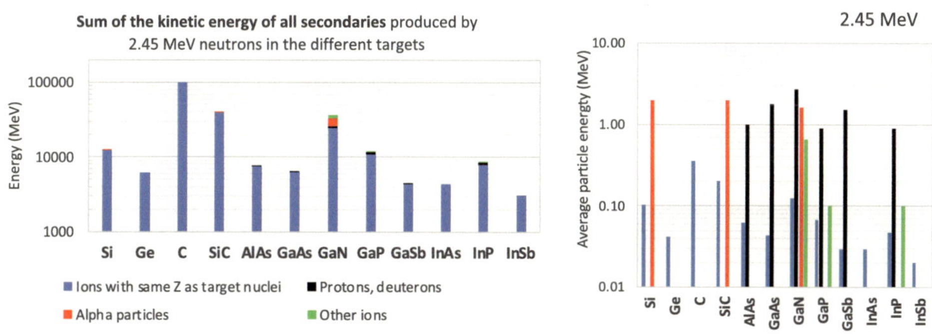

Figure 8.
Left: Sum of the kinetic energy of all secondaries produced by 2.45 MeV in the different targets (charged particle with Z ≥ 1). Right: Average particle energy per particle type (same legend as for the left figure).

significant number of nonelastic reaction channels is opened at this higher neutron energy. The ratio of the number of nonelastic reactions with respect to the total number of events varies from about 5% in the case of C up to 38% for InP; it is around 33% for Si. These nonelastic reactions produce several types of ions, protons, and alpha particles—their respective numbers are given in **Table 4** which also indicates the number of secondary ions produced characterized by the same atomic number as target nuclei. This last category includes, of course, all recoil nuclei produced in elastic and inelastic reactions but also ions produced in certain nonelastic reactions, for

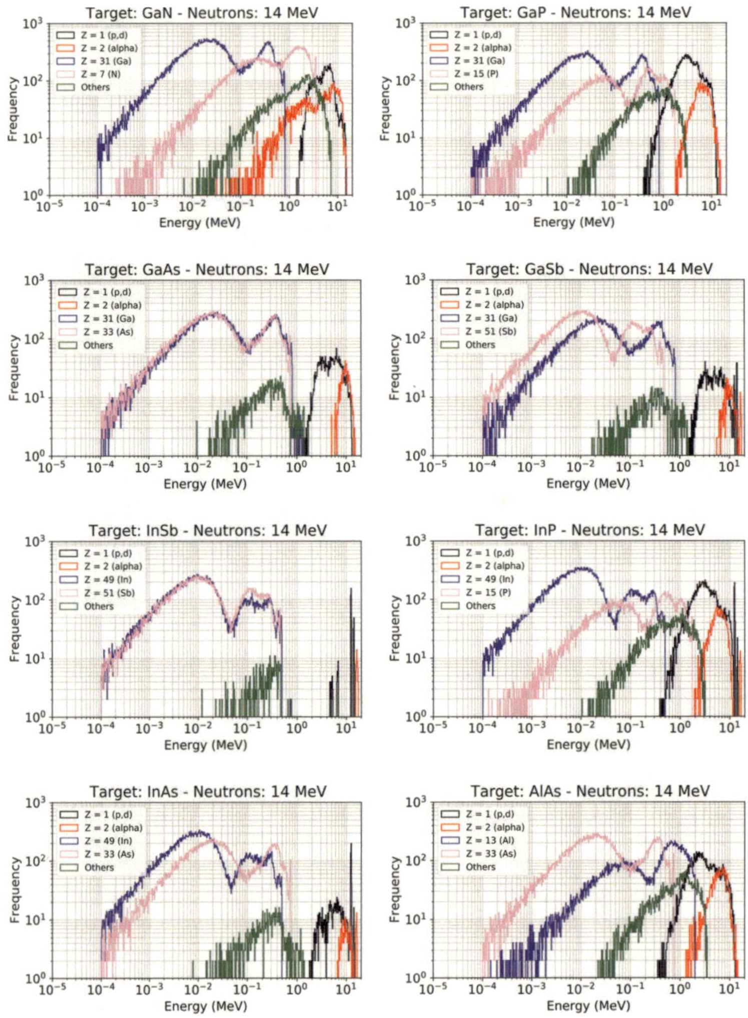

Figure 9.
Energy histograms of secondaries produced by 2.45 MeV neutrons in targets of GaN, GaP, GaAs, GaSb, InSb, InP, InAs, and AlAs III-V semiconductor materials.

example, in the (n,n'γ) or (n,2n'γ) reactions that are increasingly frequent when increasing the atomic number of the target element. These nonelastic reactions give a final nucleus with a lighter mass with respect to the initial impacted nucleus (for example, ^{74}Ge(n,2n)^{73}Ge or ^{115}In(n,2n)^{113}In, etc.).

For the good understanding of **Figures 9** and **10**, **Table 5** indicates the main nonelastic reactions leading to the production of protons and alpha particles. The energy ranges of the ejected particles are also reported. Protons, deuterons, and tritons (Z=1) are produced in large numbers in six materials, by order of importance—GaP, Si, SiC, InP, AlAs, and GaN. The main reactions for these targets are the following (we indicated in parenthesis the energy threshold of the reactions): ^{31}P(n,n'p)^{30}Si (7.5 MeV), ^{31}P(n,p)^{31}Si (0.7 MeV), ^{31}P(n,d)^{30}Si (5.2 MeV), ^{28}Si(n,p)^{28}Al (4 MeV), ^{27}Al(n,p)^{27}Mg (1.9 MeV), ^{14}N(n,d)^{13}C (5.7 MeV). Protons and deuterons are also observed in smaller numbers in the other materials, i.e., GaAs, Ge, GaSb, InAs, and InSb. Only two tritons are observed for C, they are produced in the rare reaction

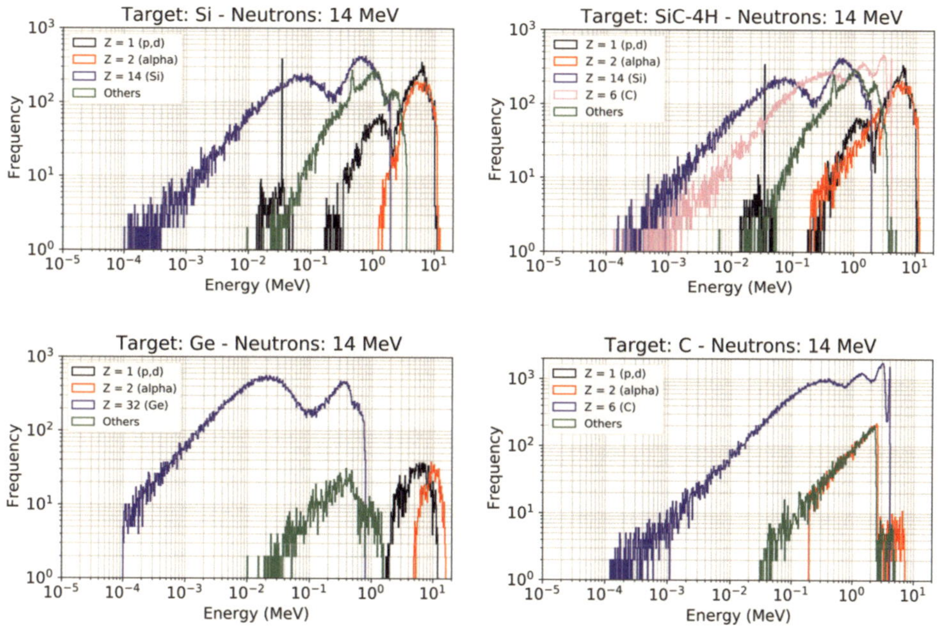

Figure 10.
Energy histograms of secondaries produced by 2.45 MeV neutrons in targets of Si, SiC-4H, Ge, and C (diamond) Group-IV semiconductor materials.

Target	Number of ions with the same Z as target nuclei	Number of protons, deuterons, and tritons	Number of alpha particles	Number of other ions
Si	58,293 (Si)	16,815	10,245	22,968
Ge	107,913 (Ge)	1910	1042	2152
C	222,165 (C)	2	11,460	11,445
SiC	56,611 (Si) + 60,415 (C)	16,376	13,101	25,577
AlAs	26,836 (Al) + 52,924 (As)	10,003	3186	5530
GaAs	51,729 (Ga) + 52,697 (As)	3238	806	1713
GaN	102,776 (Ga) + 59,190 (N)	9377	7295	12,278
GaP	57,127 (Ga) + 26,396 (P)	19,439	4126	7938
GaSb	40,309 (Ga) + 53,776 (Sb)	1874	438	1105
InAs	53,690 (In) + 42,455 (As)	1409	237	1094
InP	58,842 (In) + 20,892 (P)	13,994	2814	5725
InSb	43,067 (In) + 44,737 (Sb)	458	54	484

Table 4.
Number of secondaries produced by 14 MeV neutrons in the different targets.

$^{13}C(n,t)^{11}B$. Concerning alpha particles (Z=2), they are produced in large or significant numbers in seven materials—SiC, C, Si, GaN, GaP, InP, and AlAs. The main reactions for these targets are the following—$^{28}Si(n,\alpha)^{25}Mg$ (2.7 MeV), $^{12}C(n,\alpha)^{9}Be$ (6.2 MeV), $^{14}N(n,\alpha)^{11}B$ (0.17 MeV), $^{31}P(n,\alpha)^{28}Al$ (2 MeV), $^{27}Al(n,\alpha)^{24}Na$ (3.2 MeV). Traces of alpha particles are also observed in Ge, GaAs, GaSb, and InSb.

Element	Main reactions for α, p or d production (14 MeV neutrons)	Energy range of the ejected particle
C	$^{12}C(n,\alpha)^{9}Be$	0.2–7 MeV
N	$^{14}N(n,\alpha)^{11}B$ $^{14}N(n,d)^{13}C$ $^{14}N(n,p)^{14}C$	0.01–14 MeV 0.1–9 MeV 5–12 MeV
Al	$^{27}Al(n,\alpha)^{24}Na$ $^{27}Al(n,p)^{27}Mg$	1.8–14 MeV 0.4–13 MeV
Si	$^{28}Si(n,d)^{27}Al$ $^{28}Si(n,p)^{28}Al$ $^{28}Si(n,\alpha)^{25}Mg$	10–50 keV (peak at 34 keV) 2–10 MeV 0.5–13 MeV
P	$^{31}P(n,\alpha)^{28}Al$ $^{31}P(n,n'p)^{30}Si$	0.2–12 MeV 0.4–11 MeV
Ga	$^{69}Ga(n,\alpha)^{66}Cu$ $^{69}Ga(n,p)^{69}Zn$	4–14 MeV 0.01–14 MeV
Ge	$^{70}Ge(n,\alpha)^{67}Zn$ $^{70}Ge(n,p)^{70}Ga$	4–14 MeV 1.5–13 MeV
As	$^{75}As(n,\alpha)^{72}Ga$ $^{75}As(n,p)^{75}Ge$	4–14 MeV 1–13 MeV
In	$^{115}In(n,p)^{115}Cd$	10–14 MeV
Sb	$^{123}Sb(n,p)^{123}Sn$	12–14 MeV

Table 5.
Main reactions for proton, deuteron, or alpha particle production in the nonelastic interactions of 14 MeV neutrons with target elements. The energy ranges of the ejected particles are also indicated (see ***Figures 9*** *and* ***10*** *for visualization).*

Figure 11 (left) shows the sum of the kinetic energy of all secondaries produced by 14 MeV in the different targets. **Figure 11** (right) gives the average energy conveyed per particle for the four categories of secondaries. Protons and alphas are of prime importance in the radiation response of these materials insofar as they convey a very important part of the total kinetic energy; this part is predominant in the case of five materials—Si, SiC, AlAs, GaN, and GaP. As compared to **Figure 8** (left) obtained for 2.45 MeV neutrons, the amplification factor of the total kinetic energy at 14 MeV for

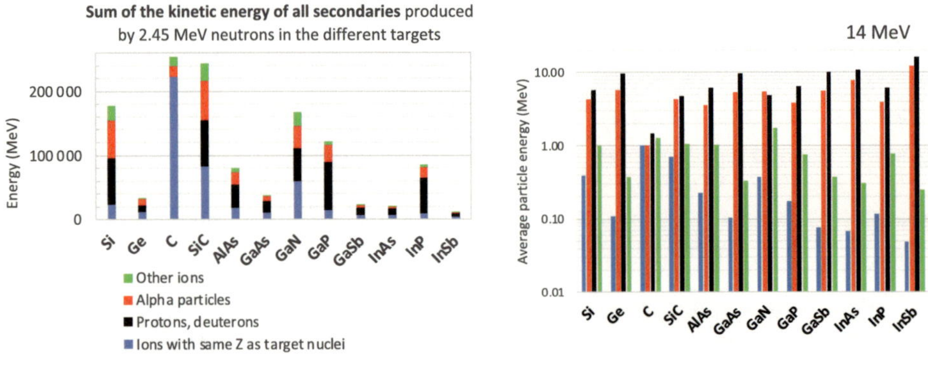

Figure 11.
Left: Sum of the kinetic energy of all secondaries produced by 14 MeV in the different targets (charged particle with $Z \geq 1$*). Right: Average particle energy per particle type (same legend as for the left figure).*

all secondaries is minimal for C (×2.5) and InSb ((×3.5) and maximal for InP, GaP, AlAs (around ×10), and Si (×14.3).

5. Implication for electronics subjected to fusion neutrons

In this last section, we examine the previous results in light of the mechanisms of creation of single-event effects (SEEs) in electronics subjected to fusion neutrons. We recall here that these SEEs are initiated by neutron-matter interactions in four main steps—(1) interactions produce secondary fragments or recoil nuclei from target nuclei. (2) The produced ion(s) interact(s) with material to generate free charge carriers via the production of electron and hole pairs. (3) Electrons and holes are transported by drift-diffusion mechanisms through the circuit materials (oxides and semiconductors) up to a sensitive node (reversely biased junctions); free carriers also recombine during their transport. (4) The collected charge on a sensitive node can significantly alter its voltage that leads to a change in device or circuit operation. This voltage glitch may propagate through the circuit. The present work allows us to provide quantitative information on steps 1 and 2 in so far as from step 3, it would be necessary to consider the circuit architecture, the applied polarizations, and the transport and recombination properties of the materials. In the following, we examine the different materials in light of the number of interactions and of the number of e-h pairs created on average per interaction.

Concerning the first metric, the number of SEEs susceptible to be created in an electronic circuit (of the same geometry as the target for simplicity) is at most equal to the number of interaction events that can deposit enough energy, i.e., create a sufficient electrical charge able to disturb the circuit operation (in the case of an SRAM memory, this minimum amount of charge susceptible to upset a memory cell is called a critical charge). This number of SEEs is not easy to evaluate but it can be upper bounded by the number of events that can deposit energy and therefore generate an electrical charge greater than or equal to the critical charge of the circuit. To roughly estimate this metric from the histogram of the energies/charges deposited per event, a simple transformation is necessary, as illustrated in **Figure 12. Figure 12** (left) shows the energy histograms (frequency and cumulative frequency histograms) for all the events produced by 2.45 MeV and 14 MeV neutrons in GaP. The relationship between energy (bottom scale) and charge (top scale) considers the energy value of the creation of electron-hole pairs, as given in **Table 2**.

At fixed energy or charge, the cumulative frequency histogram directly indicates the number of events for which the total energy deposited is below this value. In practice, it is more convenient to have the complementary value, i.e., the number of events above this value is plotted in **Figure 12** (right), and directly expressed in electrical charge instead of energy. Applying this transformation to all targets gives the ensemble of cumulative histograms, as shown in **Figures 13** and **14**.

These figures can be directly used to estimate the number of events able to deposit an electrical charge superior to a given value. From this data and knowing a minimum of geometric and electrical characteristics of the circuit to be evaluated from a radiative point of view (number and geometry of the sensitive volumes, critical charge, etc.), it would be possible, in future studies, to estimate the number of events susceptible to produce SEEs at both 2.45 and 14 MeV.

Figures 13 and **14** provide also other interesting information about the difference in the neutron susceptibility of a given material at these two energies—the greater this

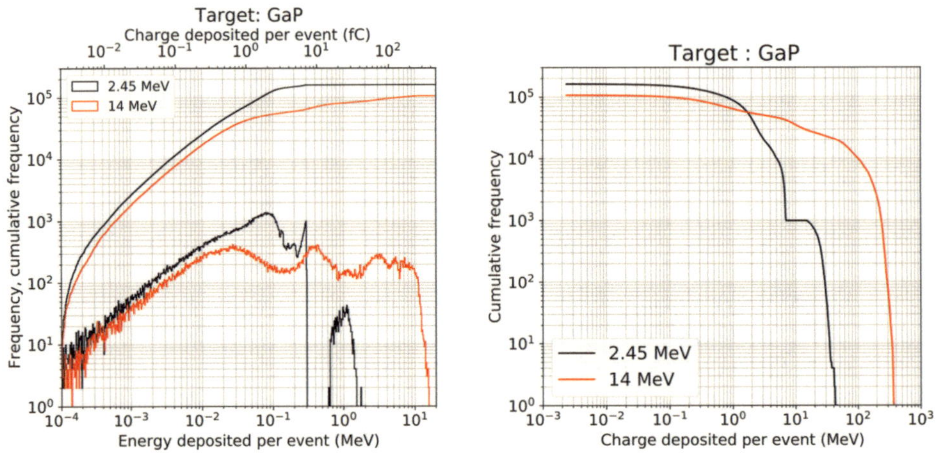

Figure 12.
Left: Histogram and cumulative histogram of the energy deposited per event for all interaction events in GaP target. Right: Reversed cumulative histogram from the left figure expressed in charge deposited per event (MeV).

difference, the more the material, and therefore, the circuit based on this semiconductor will be likely to present a significant difference in its radiative SEE response. This difference can be appreciated following two criteria—the respective amounts of events at the lowest charge deposited per event (here at 10^{-2} fC) and the "distance" between the shoulders of the curves at the highest charge values. For example, a circuit based on diamond or GaN semiconductor may logically presents less difference between its SEE responses at 2.45 MeV and 14 MeV than a circuit based on germanium, InSb, or InAs. Of course, this difference in neutron sensitivity is also dependent on the critical charge value of the considered circuit. In the case of GaAs, for example, a circuit with a 10 fC of critical charge should be quasi-immune to

2.45 MeV neutrons but sensitive to 14 MeV. To go further in this analysis and be quantitative, it would be of course necessary to consider the cumulative histograms of the charge collected at each event. The passage from the deposited charge to the collected charge requires numerical simulations on a given architecture and modeling of transport, recombination, and collection mechanisms, a work currently in progress.

Concerning the second metric, **Figure 15** gives the number of e-h pairs created in average per interaction for the different targets at the two neutron energies. At 2.45 MeV, this value is minimum for AlAs (78,000) and maximum for diamond (29,000) and Si (27,000). The average level and the relative low dispersion of these values indicate that recoil nuclei, very widely produced in these elastic and inelastic reactions, deposit small amounts of charges over very short distances of a few nanometers to a few hundred nanometers (data not shown). At 14 MeV, the results of **Figure 15** (right) show higher values (approximately $\times 3$ to $\times$ 12 for the different materials, and $\times$ 20 for Si) due to the contributions of nonelastic reactions that produce the most energetic secondaries, notably protons and alpha particles. Surprisingly and contrary to the case of irradiation with atmospheric neutrons [10], Silicon sensibly differs from the other materials at 14 MeV since it has the greatest multiplying factor between

2.45 and 14 MeV. This is due to a relatively low number of events (86 kilo events) at this energy with respect to the other semiconductors (**Table 6**) combined with an important level of production of protons and alpha particles (**Table 3**) and also with an energy value of e-h pair creation inferior to the average value (around 5 eV) for all

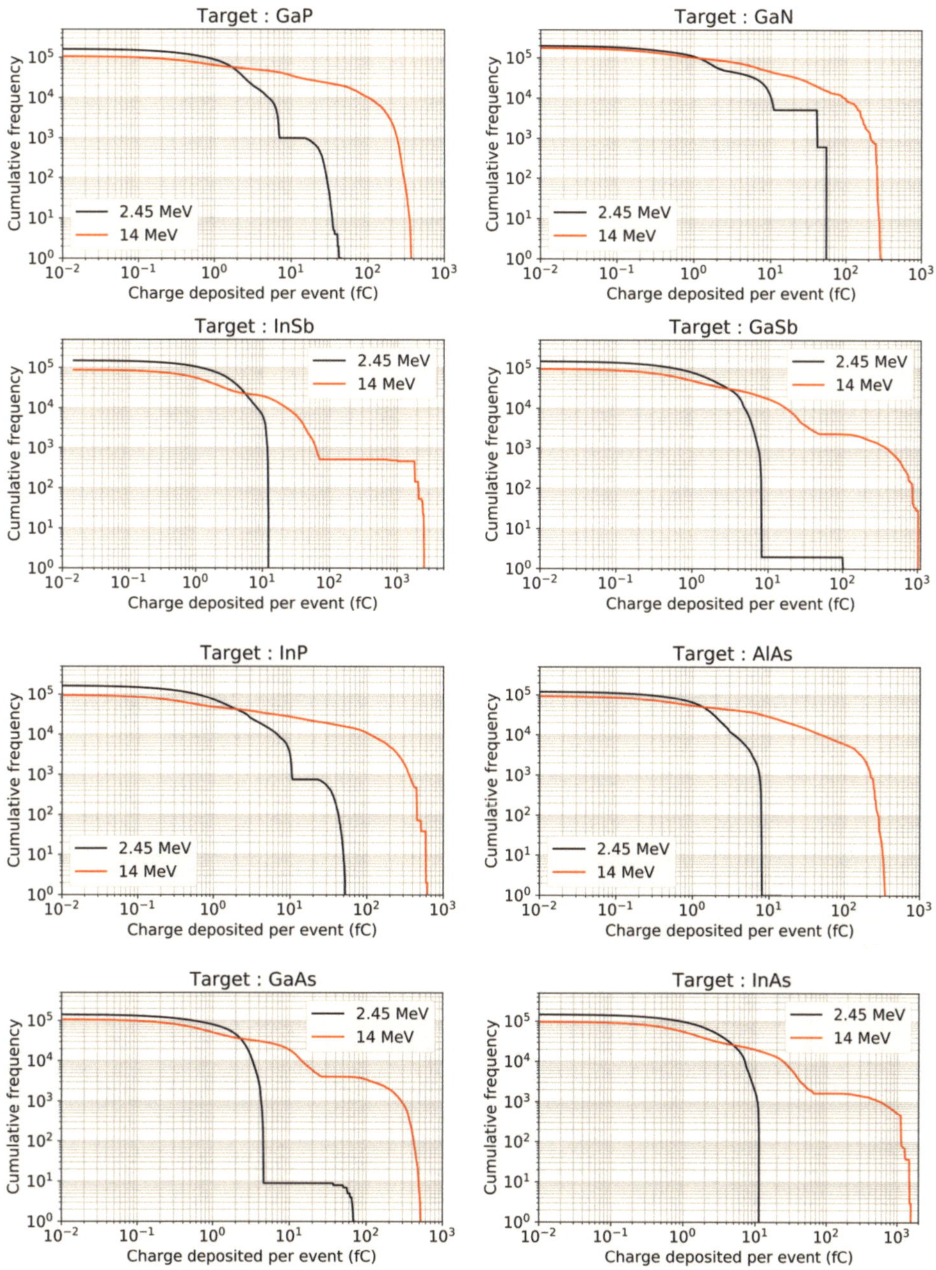

Figure 13.
Reversed cumulative histogram of the energy deposited per event for all interaction events in GaP, GaN, InSb, GaSb, InP, AlAs, GaAs, and InAs targets subjected to 2.45 MeV and 14 MeV neutrons. For a fixed value of the deposited charge, the curves give the number of interactions able to deposit at most this energy in the corresponding target.

materials (**Table 2**). The other materials which exhibit high values are SiC, GaP, and InP. Statistically, in these four materials, a neutron-target interaction at 14 MeV will deposit more charge than in the other materials, which potentially makes these events more threatening for the creation of SEEs. The balance between all these factors is, therefore, not obvious when one remains at the only "material" level for these studies.

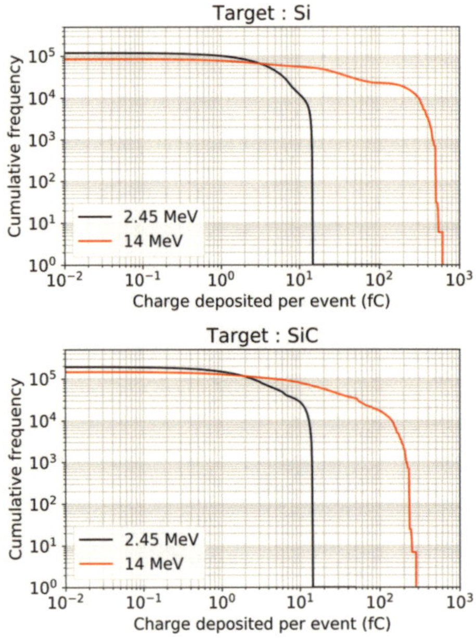

Figure 14.
Reversed cumulative histogram of the energy deposited per event for all interaction events in Si, Ge, SiC, and C (diamond) targets subjected to 2.45 MeV and 14 meV neutrons. For a fixed value of the deposited charge, the curves give the number of interactions able to deposit at most this energy in the corresponding target.

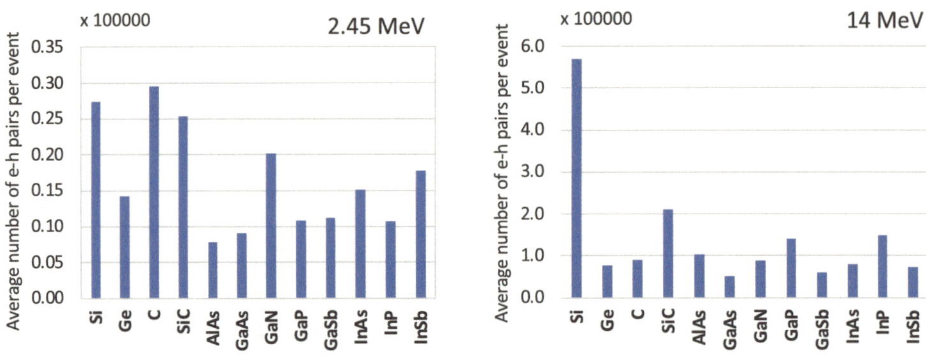

Figure 15.
The average number of e-h pairs created per interaction event in the different targets subjected to 2.45 MeV (left) and 14 MeV (right) neutrons.

This point requires numerical simulations at the circuit level to go further in the investigation of the level of SEE in electronics based on these different materials.

6. Conclusion

This work explored the radiation response of Group IV (Si, Ge, SiC, diamond) and III-V (GaAs, GaN, GaP, GaSb, InAs, InP, InSb, AlAs) semiconductors subjected to D–D (2.45 MeV) and D–T (14 MeV) neutrons. This first study was limited to the level of the neutron response of bulk materials, an essential step to be able, subsequently, to study the response of electronic components and circuits based on these materials. The response of each semiconductor has been systematically investigated through a direct calculation using nuclear cross-section libraries, MCNP6, and Geant4 numerical simulations. The counting of reaction rates per type of reaction (elastic, inelastic, nonelastic), as well as the classification of all neutron-induced secondary products as a function of their nature and energy, gives a first very complete picture of the behavior of these materials in a fusion environment under D–D or D–T neutrons. Although all studied materials exhibit a larger total number of interaction events than silicon, both at 2.45 and 14 MeV, there is nothing to conclude at this stage that they will present a larger response in terms of SEEs at circuit level since this higher neutron susceptibility only relates to the overall number of events and not to the production of the most energetic particles, such as protons or alpha particles. Moreover, such a higher neutron response may be compensated by other mechanisms at the electronics level, for example, a higher energy value of e-h pair creation (which has the effect of reducing the charge deposited) or higher carrier mobility (which can increase the performance of transistors and strengthen the resilience of circuits to single events). Future studies at the circuit level, following a methodology similar to a first study carried out on GaN subjected to atmospheric neutrons [19], will be largely based on the interaction event databases compiled during this first work and will provide more quantitative information to precisely assess the D–D and D–T neutron radiation response of future circuits based on these alternative materials to the classical silicon of microelectronics.

Conflict of interest

The authors declare no conflict of interest.

Appendix

Target (natural material)	Library or simulation code	Number of interaction events (1cm^2 $\times$ 20 μm, 5$\times$10^8 neutrons)					
		2.45 MeV			14 MeV		
		Elastic	Inelastic	Nonelastic	Elastic	Inelastic	Nonelastic
Si	TENDL-2021	94,285	31,604	0	37,776	24,611	33,529
	Geant4.10.07	92,927	27,897	4	34,665	23,623	28,341
	MCNP6.2	94,098	30,426	22	33,452	26,309	28,547
Ge	TENDL-2021	77,667	70,778	1456	76,038	32,294	38,127
	Geant4.10.07	77,565	70,315	305	77,241	30,662	40,281
	MCNP6.2	80,378	70,084	323	79,224	24,639	43,772
C	TENDL-2021	284,170	0	0	147,136	74,624	11,862
	Geant4.10.07	283,451	0	0	146,732	75,433	11,462
	MCNP6.2	282,482	0	0	142,767	74,664	14,726
SiC-4H	TENDL-2021	168,715	30,466	0	76,711	44,162	35,570
	Geant4.10.07	167,221	26,879	7	73,700	43,317	30,745
	MCNP6.2	168,151	29,333	20	71,230	45,704	31,559
AlAs	TENDL-2021	71,069	72,328	1570	56,868	23,143	33,523
	Geant4.10.07	70,519	50,256	208	57,846	21,907	34,300
	MCNP6.2	71,557	53,096	194	55,500	23,968	35,153

Target (natural material)	Library or simulation code	Number of interaction events (1cm^2 × 20 μm, 5×10^8 neutrons)					
		2.45 MeV			14 MeV		
		Elastic	Inelastic	Nonelastic	Elastic	Inelastic	Nonelastic
GaAs	TENDL-2021	69,858	72,905	334	76,848	26,038	44,225
	Geant4.10.07	69,858	72,905	334	78,127	26,069	43,719
	MCNP6.2	74,428	71,515	354	73,757	30,081	43,452
GaN	TENDL-2021	129,877	66,514	7683	117,421	42,330	54,074
	Geant4.10.07	130,978	65,011	5420	119,180	42,761	52,815
	MCNP6.2	138,249	55,632	5573	114,349	47,939	50,588
GaP	TENDL-2021	109,534	50,357	1317	60,171	21,781	45,557
	Geant4.10.07	111,630	49,735	1111	60,853	22,646	44,911
	MCNP6.2	98,073	46,664	1309	64,006	32,412	30,478
GaSb	TENDL-2021	90,943	54,223	1189	77,215	17,192	45,596
	Geant4.10.07	90,310	57,992	1072	72,605	21,477	46,037
	MCNP6.2	101,262	46,366	1026	71,519	26,181	42,512
InAs	TENDL-2021	85,006	71,752	4280	78,268	18,035	47,203
	Geant4.10.07	82,040	64,307	2007	78,346	17,799	46,801
	MCNP6.2	88,362	63,776	1824	83,978	15,828	45,441
InP	TENDL-2021	117,899	54,714	4370	65,719	14,025	48,965
	Geant4.10.07	118,171	45,215	2689	65,177	14,538	48,000
	MCNP6.2	109,501	43,783	2720	77,879	16,575	35,607
InSb	TENDL-2021	98,085	55,811	3451	77,290	11,768	47,215
	Geant4.10.07	96,303	52,707	2264	72,622	15,152	47,707
	MCNP6.2	107,126	43,471	2080	79,225	14,813	43,781

Table 6.
*Details of numerical values for **Figures 2–4**.*

Author details

Jean-Luc Autran* and Daniela Munteanu
Aix-Marseille University, CNRS, University of Toulon, Faculté des Sciences, Marseille Cedex, France

*Address all correspondence to: daniela.munteanu@univ-amu.fr

References

[1] ITER. Unlimited Energy [online]. Available from: http://www.iter.org

[2] Lee S. Advanced Material and Device Applications with Germanium. London: IntechOpen; 2017

[3] Kimoto T, Cooper JA. Fundamentals of Silicon Carbide Technology: Growth, Characterization, Devices, and Applications. Singapore: John Wiley & Sons; 2014

[4] Oktyabrsky S, Ye P. Fundamentals of III-V Semiconductor MOSFETs. Berlin: Springer Science & Business Media; 2010

[5] Li T, Mastro M, Dadgar A. III–V Compound Semiconductors: Integration with Silicon-Based Microelectronics. Boca Raton: CRC Press; 2016

[6] Wesson J. Tokamaks. 4th ed. Oxford: Oxford Science Publication; 2011

[7] Morse E. Nuclear Fusion. Cham: Springer Nature Switzerland AG; 2018

[8] Munteanu D, Autran JL. Susceptibility of group-IV and III-V semiconductor-based electronics to atmospheric neutrons explored by geant4 numerical simulations. In: Rao SP, editor. Numerical Simulations in Engineering and Science. London: IntechOpen; 2017

[9] Munteanu D, Autran JL. Interactions between Terrestrial Cosmic-Ray Neutrons and III-V Compound Semiconductors. In: Valdman J, Marcinkowski L, editors. Modeling and Simulation in Engineering – Selected Problems, Edited by. London: IntechOpen; 2020

[10] Autran JL, Munteanu D. Atmospheric neutron radiation response of III-V binary compound semiconductors. IEEE Transactions on Nuclear Science. 2020;**67**:1428-1435

[11] TENDL-2021. TALYS-Based Evaluated Nuclear Data Library [online]. Available from: https://tendl.web.psi.ch/tendl_2021/tendl2021.html

[12] ENDF Manual. Appendix B: Definition of Reaction Types [online]. Available from: https://www.oecd-nea.org/dbdata/data/manual-endf/endf102_MT.pdf

[13] Agostinelli S et al. Geant4 – A simulation toolkit. Nuclear Instruments and Methods in Physics Research Section A: Accelerators, Spectrometers, Detectors and Associated Equipment. 2003;**506**(3):250-303

[14] Allison J, Amako K, Apostolakis J, Arce P, Asai M, et al. Recent developments in Geant4. Nuclear Instruments and Methods in Physics Research Section A: Accelerators, Spectrometers, Detectors and Associated Equipment. 2016;**835**:186-225

[15] Geant4, Geant4 10.7 Release Notes [online]. Available from: https://geant4-data.web.cern.ch/ReleaseNotes/ReleaseNotes4.10.7.html

[16] Werner CJ et al. MCNP6.2 release notes. Los Alamos National Laboratory, report LA-UR 18-20808. 2018

[17] C.J. Werner (editor), MCNP User's Manual – Code Version 6.2, Los Alamos National Laboratory, report LA-UR 17-29981, 2017

[18] Autran JL, Munteanu D. Soft Errors: From Particles to Circuits. Boca Raton (FL), USA: Taylor & Francis/CRC Press; 2015

[19] Munteanu D, Autran JL. Terrestrial neutron-induced single events in GaN. Microelectronics Reliability. 2019;**100**: 113357